Ergonomics for the Layman

Ergonomics for the Layman

Applications in Design

Prabir Mukhopadhyay, MSc, PhD

CRC Press
Taylor & Francis Group
Boca Raton London New York

CRC Press is an imprint of the
Taylor & Francis Group, an informa business

CRC Press
Taylor & Francis Group
6000 Broken Sound Parkway NW, Suite 300
Boca Raton, FL 33487-2742

Printed on acid-free paper

International Standard Book Number-13: 978-0-367-33499-4 (Hardback)

Library of Congress Cataloging-in-Publication Data
Names: Mukhopadhyay, Prabir, 1969- author.
Title: Ergonomics for the layman : applications in design / by
Prabir Mukhopadhyay.
Description: Boca Raton, FL : CRC Press/Taylor & Francis Group, 2019. |
Includes bibliographical references and index.
Identifiers: LCCN 2019020292 | ISBN 9780367334994 (hardback : acid-free paper) |
ISBN 9780429321627 (ebook)
Subjects: LCSH: Human engineering–Popular works.
Classification: LCC T59.7 M85 2019 | DDC 620.8/2–dc23
LC record available at https://lccn.loc.gov/2019020292

Visit the Taylor & Francis Web site at
www.taylorandfrancis.com

and the CRC Press Web site at
www.crcpress.com

MIX
Paper from
responsible sources
FSC® C013056

Printed and bound in Great Britain by
TJ International Ltd, Padstow, Cornwall

Dedication

To the Lotus Feet of Sri Ramakrishna Paramhansha Dev for whom this impossible mission was possible, my deepest regards to you.

To Bapi (my father, Late Dhirendra Nath Mukhopadhyay), for all your love ... miss you a lot!

To Ma (my mother, Meena who still sees the child in me and feels that I should never give up), for all your love ... keep praying for us Ma.

To Dada (my brother, Prodyut, for all his unconditional love and affection), love you, Dada.

To Mamoni (my sister-in-law, Juin, for looking at me as the future family custodian), keep smiling forever, Mamoni.

To Ghantu (my niece, Aritriya) and Leto (my nephew, Pratik), for being my best buddies ... love you, folks.

To my two great mentors, the late Prof. R.N. Sen, Calcutta University, India, my postgraduate supervisor for introducing me to the "world of ergonomics"; and Prof. T.J. Gallwey, University of Limerick, Ireland, my PhD supervisor, for teaching me what miracle ergonomics can do. My deepest regards to you both.

To Dr. S. Ghosal, my senior colleague and mentor at the National Institute of Design, India, for teaching me how to "teach" ergonomics to the budding designers. Thank you so much, Soumyajit Da.

And to all those who wished to see me successful and whose names I might have missed. I owe you all.

Contents

Preface

There has been a long-standing demand from my design students, mostly from those who do not have any background in science and technology, for a book on ergonomics that they could read, understand, and apply in their designs. This is an attempt in that direction. This book is in no way a replacement for other advanced books on ergonomics in the market.

The book has been written in simple, easy to understand language, so that any layman will be able to understand it. The book opens with an introduction to the subject, gradually introducing the readers to different broad topics such as systems and dimensions, and then touches on different domains like products, space, and communication. The purpose is to let readers understand the potential of ergonomics in different areas of design. Each chapter is followed by assignments for the students to test their understanding of the content of the chapter. A separate chapter contains more detailed assignments with guidelines to solving them. This section should be undertaken once all the chapters have been covered and understood.

The language of the book is in a story telling format, the way I prefer teaching the subject in the classroom. This is to ensure that the subject does not become monotonous for the design students and so that they develop the curiosity to know more. The images used in this book are for representational purposes only, and do not contain any technical details. The purpose is to build a story line and ensure that the reader is never bored while reading the text.

I am grateful to many people for inspiring me to write this book. I would like to remain in the hearts of my students through this book when I am no longer in this world. The idea of this book was conceived at the beautiful lake city, Udaipur in India, the venue for the famous James Bond movie *Octopussy*. There my friend and colleague Dr. Shekhar Chatterjee inspired me to write a book on ergonomics so that every designer could understand and use it. Had it not been for his inspiration on that beautiful moonlit night on the banks of Lake Pichola, the book would have only been a dream. All my students have repeatedly requested that I write a book in the manner I teach in the classroom, so that it's easy to understand and apply. Again, I wish to express my gratitude to those people who have helped in preparing this book. I am particularly grateful to Ms. Erin Harris, Mrs. Cindy Renee Carelli, and the staff of CRC Press/Taylor & Francis Group, for their continued editorial and production support throughout the publication process.

Acknowledgments

Mr. Vipul Vinzuda, Associate Senior Faculty, Transportation and Automobile Design, National Institute of Design, Gandhinagar, India, who did all the illustrations for this book.

The feedback of all the reviewers of the book is sincerely acknowledged.

Some of the photographs used in the book have been taken from royalty-free websites. I am grateful to all the photographers for sharing their photographs for free with all. Thank you so much.

Author

Prabir Mukhopadhyay, MSc, PhD, holds a BSc Honours degree in physiology and an MSc in physiology with a specialization in ergonomics and work physiology, both from Calcutta University, India. He holds a PhD in industrial ergonomics from the University of Limerick, Ireland. Dr. Mukhopadhyay started working with noted ergonomist Prof. R.N. Sen at Calcutta University for both his master's thesis and later on a project sponsored by the Ministry of Environment and Forests, Government of India. It was during this time that he developed a keen interest in the subject and decided to build a career in ergonomics. He joined the National Institute of Design, Ahmedabad, India as an ergonomist for a project for the Indian Railways. There, he was mentored by Dr. S. Ghosal, the project lead. He then joined the same institute as a faculty in ergonomics. During his tenure at Ahmedabad, he worked on many consultancy projects related to ergonomics. Some of his clients included the Indian Railways, Self-Employed Women's Association, and the United Nations Industrial Development Organization.

After two years, Dr. Mukhopadhyay left for the University of Limerick, Ireland, on a European Union–funded project under the supervision of Prof. T.J. Gallwey. He completed his PhD in industrial ergonomics at the same university, and decided to return to India to apply his acquired knowledge. He joined the National Institute of Design, Post Graduate Campus at Gandhinagar, India, as a faculty in ergonomics. There he headed the Software and User Interface Design discipline. He also completed a research project funded by the Ford Foundation – National Institute of Design on ergonomics design intervention in the craft sectors at Jaipur in Rajasthan, India. Simultaneously, he started teaching ergonomics across different design disciplines at other campuses of the institute, such as Ahmedabad and Bangalore.

Five years later, Dr. Mukhopadhyay joined the Indian Institute of Information Technology Design and Manufacturing Jabalpur, India as an assistant professor in design. He was then promoted to associate professor, and later became the discipline head. He teaches, practices, and researches in different areas of ergonomics and its applications in design. He is a bachelor, and his hobbies include watching action movies, listening to Indian and Western music, traveling, and cooking. He may be reached at prabirdr@gmail.com.

1 Introduction

OVERVIEW

This chapter introduces readers to the concept of ergonomics and its pertinence to design. The genesis of ergonomics, including its application areas, is outlined in this chapter. At the end of the chapter, readers are expected to have a broad understanding of ergonomics and how it can be used in their respective fields of design.

1.1 ERGONOMICS

Ergonomics is the relationship between human, product, and the environment in which the human exists. For example, let us imagine that a human is breaking a piece of coal into smaller pieces. The components involved are the human, the hammer, and the platform on which the coal is kept and broken. The environment could be outdoors or indoors, wherever the activity is taking place. Now let us consider the factors that could affect the human's productivity, or that could enable him to break the maximum amount of coal into smaller pieces in the shortest span of time. The answer is not very straightforward, but demands a holistic view of the situation. One needs to look at the different components that we talked before. Let's take the human first; he has to be healthy and trained to do the job – only then he can do it effectively. So, the human's health condition also affects productivity. The second element is the hammer, especially the design of the handle of the hammer. If the hammer handle is too short it will be difficult to hold and hence to strike with the required force. If the handle is too long it becomes difficult to grip and force exertion becomes difficult. Thus, hammer handle design is an important element that can affect productivity. The third element is the environment in which the person is working. If the work is done outdoors under the scorching sun, the person will be more tired and his productivity will be less compared to if the person works indoors in a relatively cooler environment. So, the environment also affects productivity. Can anything else affect productivity? Yes, there is probably another factor which affects productivity, and that is the psychological factor. Let's imagine that the person's son is ill at home. So as a dad he would be worried about his son and this might also affect his performance or the overall productivity. Thus, a simple task of breaking a piece of coal has so much complexity involved in it, and each smaller component has a decisive role to play when it comes to performance and productivity.

A small thorn got stuck in my left finger early one morning when I was picking flowers in the garden. In an attempt to take the thorn out of my fingers (Figure 1.1), I first pressed the left fingers with my right fingers, but with no success. Next I tried to take the thorn out of my fingers with my nails, thinking that they were relatively sharper and would help. It did not help either. Then I took a needle, used the tip of it to take the thorn out of my finger successfully. If you analyze the stages involved in removal of the thorn from my finger then you would notice a sequence of activities.

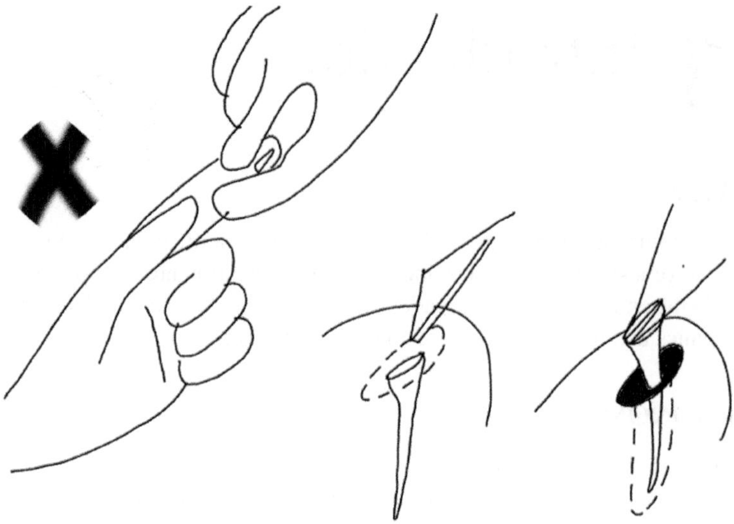

FIGURE 1.1 Ways of removing a thorn stuck in the finger.

First, I tried by pressing the finger (a rather crude approach) then I tried using my nails (a relatively finer approach), and finally I used a needle. In essence, the two examples of what happened is that the tool or the needle acted as an extension of the human hand and amplified human capacity. This is exactly what ergonomics is and where it plays a role in design. It acts at the junction between human and product/space or communication, and so on, considers the different issues surrounding it and then helps in designing the process so as to increase productivity, comfort, well-being, occupational health, and safety of the users. Thus, ergonomics makes the design much more user friendly or humane, because tools, products, and machines are for the benefit of the human, to amplify his capabilities.

Ergonomics comprises two broad categories: physical ergonomics and cognitive ergonomics. Physical ergonomics is that tangible part of ergonomics which can be felt, measured directly. Cognitive ergonomics is the softer side of ergonomics – the person thinking about his ailing child at home as in the example above – and cannot be felt, or measured directly. Although these two areas of ergonomics are inseparable, we will deal with them separately for the ease of our understanding of the subject matter.

1.2 GENESIS

Ergonomics as a multidisciplinary subject draws from science, technology, psychology, anatomy, physiology, and so on. It has existed ever since humans started hunting and gathering food. For these activities, the human developed his own tools. This was made out of wood and stone. In designing these tools for survival, the human took reference of his own body parts or those of the family members who were residing with him. This was possible through repeated trial and error. This was the first application of ergonomics in primitive times by humans. After the Industrial Revolution, the prime purpose of the human changed. From hunting and gathering food for himself, the human focused on mass-scale manufacturing, for himself and for others. Unfortunately, the structure of the human body did not evolve on a par with that of rapid advancement of technology, even till now. Due to large-scale manufacturing, the human started making products no longer for himself alone, but also for others, and the products gradually increased in number. It was at this juncture that there was a need for a standardization of tools and equipment, as people varied among themselves not only in terms of body dimensions but also in terms of strength, physique, and intelligence. Thus, the need for standardization of tools came to the forefront, and humans first felt the need of ergonomics in the design process. This was all unofficial and not well documented.

The first official genesis of ergonomics took place during some wars. At the time, it was noticed that trained pilots were making mistakes in activating the right controls when needed. The defense experts perceived that good aircraft with trained pilots would win battles. This did not happen. When investigated with the pilot of one of the aircrafts, an interesting aspect came forward. The pilot in one of the aircraft was hit by the enemy's anti-aircraft gun, but somehow managed to escape and return to base. The pilot was narrating that everything was going fine during the mission; he was planning for dropping a bomb over the target in the enemy territory. All of a sudden, he realized (Figure 1.2) that instead of pulling the control for dropping the bomb, he had inadvertently pulled the landing gear, leading to the rapid descent of the

FIGURE 1.2 Pilot activating the wrong gear during action. (Photo by Mike Yakaites from Pexels. Accessed April 5, 2019.)

aircraft and hence coming within range of the enemy's anti-aircraft gun. Fortunately, the pilot realized his mistake and activated the correct gear, taking the aircraft out of enemy territory and returning safely to the base. This incident was a trigger to the defense experts. For the first time they realized that pilot performance inside the cockpit of an aircraft is dependent on many factors. First is the design of the cockpit, keeping the user in mind. Second is the physical environment inside the aircraft, and the third factor is the enormous amount of psychological stress that the pilot has to withstand. This led to the fact that when a machine or a piece of equipment is designed, one has to keep the user in mind and also the context in which the machine or equipment will be used. In this case, the aircraft was never designed factoring in issues of users' psychological stress, body dimensions, and the likelihood of making mistakes. The mistake was made because the two controls for releasing the bomb and landing the aircraft were similarly shaped and placed close to one another. Under tremendous stress the pilot had to look in front, and had to locate the right control and activate it to release the bomb by feeling with his hand. By mistake, his hand fell on the landing gear and the aircraft started descending. This proves once again the importance of humans and led to the concept of human, machine, and environment and the interaction between the three in designing equipment or products.

1.3 DESIGN AND ERGONOMICS

Ergonomics and design are intricately associated with one another like "blood brothers". We have seen what ergonomics is by now. Design imparts values and these values manifest in the form of tangible outputs or "products", or they may manifest in the form of intangible outputs or "processes". In both these cases, the human and the user is always involved. So, ergonomics becomes essential throughout the design process and keeps the designer on the right track, in terms of human dimensions, characteristics, limitations, desires, and values. This makes the entire design and the design process much more user centric and makes it acceptable to a wide spectrum of the population who vary enormously in terms of dimensions, strength, thought processes, and so on. In essence, ergonomics tries to "refine" the design to make it more acceptable to a large number of users by accounting for different "human elements" and ensuring that these are incorporated in the design, be it a product, space, communication, or service. This approach gives any design that "human touch", which makes you say "wow, wonderful!"

1.4 APPLICATION AREAS

Ergonomics can be divided into two broad categories: physical or tangible ergonomics, and cognitive or intangible ergonomics. Although these two divisions exist for the sake of our understanding of the subject matter, they are an integral part of one another in design. The reason for this integration will be discussed in later chapters of this book. When it comes to the application of ergonomics, it is virtually everywhere, wherever humans are involved. Starting from a hammer, the design of the handle has to be in tandem with human hand dimensions. If one has to design a large space like a classroom, then the dimensions of the students as well as the number of students have to be considered. If we move into a complex domain like the cockpit of an aircraft, there also ergonomics will dictate the design of display and control elements, the dimensions of the controls, the type and quantum of information on the display so that the pilot is at ease while flying. In essence, everywhere from pin to plane, wherever there is a human involved using the design or process, ergonomics plays a role. Let's take the example of an electronic assembly line. It is a process which can be designed. Ergonomics will play a role here by directing how long the person should work at a stretch, how many repetitive movements per minute he should perform so as to improve productivity and the quality of the job (with minimal rejections), and facilitating occupational health and safety.

1.5 ERGONOMICS IN THE ROLE OF A DOCTOR

Ergonomics plays the role of a general practitioner (GP) in the design process. When you have a fever and go to the GP, what does he do? There are certain protocols that are normally followed. First, he gives some medicine (e.g., paracetamol) to control the fever and asks you to observe for a few days (Figure 1.3). If the fever does not subside, then you are asked to go for blood and other tissue fluid tests. If the fever is still not under control, then you are asked to go for an X-ray, or a CT scan or an MRI depending upon the severity of the disease. If he senses any danger, he will refer you to a specialist. Thus, the GP has a very holistic view of your disease and orchestrates the entire treatment in a sequential manner. In design, ergonomics also plays the role of a GP. If someone is designing a hammer, the designer would probably focus

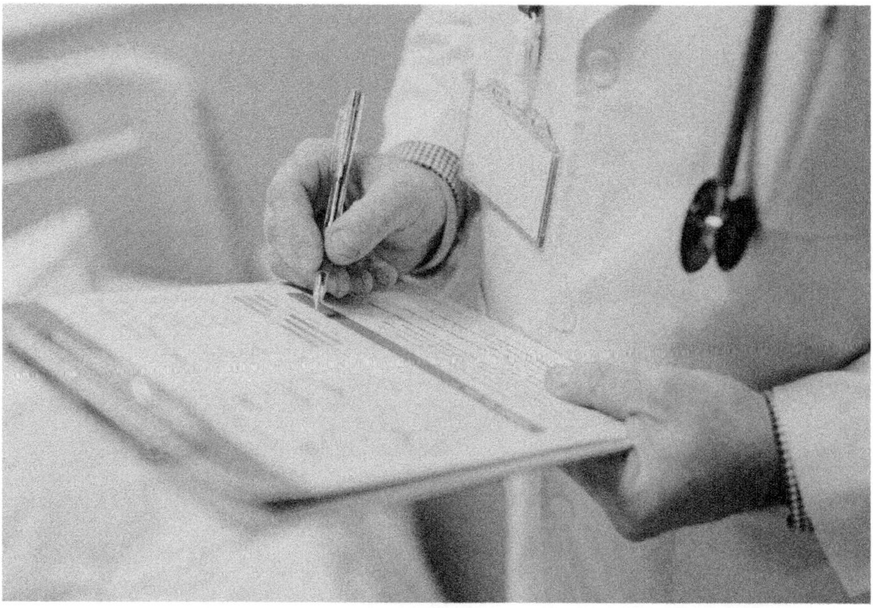

FIGURE 1.3 Ergonomist and the doctor in design. (By rawpixel.com from Pexels.)

only on the form and material of the hammer. Ergonomics will contribute to the dimensions of the hammer in tandem with human hand dimensions. The ergonomist will go further in advising the user on how to use the hammer so that he might get the maximum output from it. For example, if the hammer is used for a long time to hit a nail from overhead, that might lead to shoulder pain. That does not indicate bad hammer design, but rather bad usage of a well-designed hammer. This advice comes from an ergonomist.

1.6 DIVISIONS OF ERGONOMICS IN DESIGN

Ergonomics can be applied in the major aspects or types of design: product, space, and communication. In each of these design domains, ergonomics, both physical as well as cognitive, has an immense role to play. In the subsequent chapters we shall discuss them in greater detail.

1.7 ASSIGNMENTS

1. Pick up any product of day-to-day use like a hammer, a pencil sharpener, or a paperweight and list the different ergonomic and unergonomic issues in them. You need not worry about the solutions at this stage.
2. List some of the areas in and around you like the kitchen at home or your study table where ergonomics could be applied.
3. Try to identify ergonomic issues related to the design of a product and space around you.

2 Systems Perspective of Ergonomics in Design

OVERVIEW

This chapter introduces readers to the concept of systems ergonomics. This concept is very important for budding designers in almost all fields of design. Issues involving types of systems, reliability of systems, and why and how ergonomics is known as a systems concept are discussed in this chapter. At the end of this chapter, students will learn to think about a design problem from a more holistic perspective. The words "product" and "machine" are generic and have been used interchangeably to denote any artificial human-made object which helps a human to achieve its goal.

2.1 WHAT IS A SYSTEM?

Albert Einstein once said that humans have made this world very complex day by day. So to solve problems which are in essence complex in nature, we need to address them at that level of complexity, or else solutions would not be possible. This is exactly where we stand today with technology. We have been focusing only on developing technology for the benefit of humankind, but rarely do we think of the person (user) who is going to use the technology and the context under which this technology would be used. The resultant effect is evident; technology has failed to address many of the problems faced by humans to date. The genesis of ergonomics which we discussed in the previous chapter, wherein a fighter aircraft was being shot down by the enemy because of a pilot error, was an example of systems failure. That error was just the "symptom" of a larger problem and not the cause. It was an example of a systems-level problem where the design of the aircraft controls was at fault, and not the pilot. There are a number of such systems-level problems around us, and to solve them a similar systems-level approach is required. A systems concept (Figure 2.1) is that which visualizes a product, space, or interaction by looking at the interconnections and relationships, the entire picture as well as the smaller components making up the larger picture. It's just like a pyramid: if you look from a distance it appears as a singular solid structure, but if you look at it closely you find that it is made up smaller components that are connected to form the larger structure.

In other words, the systems concept is that which deals with more than one entity, that is two and more entities interacting among themselves for a common purpose. Hence ergonomics is also a systems concept because there are three major elements: human, product, and environment, and all three interact with one another for a common purpose (thus ergonomics and systems are blood brothers!). There are

FIGURE 2.1 Systems concept in ergonomics. (Photo by Mustafa Ezz from Pexels.)

such examples of systems around us. Surgeons operating on a patient inside an operating theater are also the part of a larger system. The hospital has its infrastructure. Infrastructure alone would not work, so supporting staff, other doctors and nurses, are required. Each component of the hospital system is interrelated with for the well-being of a patient. Doctors interact directly with patients. Supporting staff interact with patients and patient family. Doctors, on the other hand, interact directly with equipment. So in this complex system if any one of the components is removed then patient well-being is affected. For example, if housekeeping and front office staff go on strike, then there would be no housekeeping and no one to interact with the patient and their family, so the entire work of the hospital would be at stake. Ergonomics from a systems perspective teaches the designer to look into any design in a context and never as a mere artifact. The role of a designer goes far beyond designing product and spaces by incorporating the human touch into the design, which is achievable by following the systems concept, thus ensuring that the design is much more enriched in terms of its usefulness and usage by humans.

2.2 OBJECTIVE OF SYSTEMS ERGONOMICS

The main objectives of systems ergonomics in the design process and overall design are increasing human efficiency with the product or space and to increase desirable human values. In other words this "way of systematic thinking" gives an insight into the management of complexity by applying ergonomic principles. This objective of systems ergonomics mainly focuses two directions: mission oriented

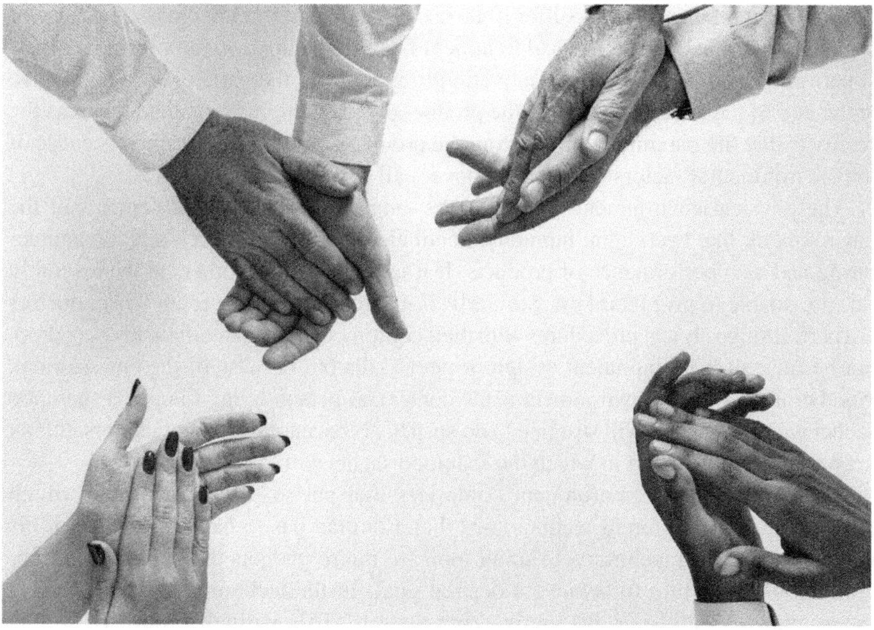

FIGURE 2.2 Systems ergonomics orientation: Mission and service. (Photos by rawpixel.com from Pexels.)

and service oriented (Figure 2.2). Mission-oriented systems ergonomics is where the target is a particular mission, for example to rescue people in danger after an earthquake. Service-oriented systems ergonomics is where a service is provided. A hotel is the best example, where the prime target is to provide quality service to the guests.

2.3 PARTS OF SYSTEMS ERGONOMICS

The systems perspective of ergonomics comprises different parts. First is the environment (heat, humidity, illumination, etc.) or the context (wearing gloves and operating a smartphone, trying to open a tap with wet hands, driving through torrential rainfall on a slippery road, etc.) where the work is being done. This is followed by the technology or design in context and then the psychological (mental attributes) and social (belief, politics, etc.) aspects of the environment.

The work environment entails not only that mentioned above, but also the environment where the work takes place along with the components of the surroundings, which are not necessarily always tangible. For example, it deals with the organizational structure including hierarchy, relationship of employees with their colleagues, superiors, and subordinates. It also deals with the power and decision-making capability of each individual, all of which directly or indirectly affect the design or the design process if it is in the manufacturing sector. The performance of the product or space or interaction might suffer if the user is under stress or depressed because of organizational factors such as bad treatment from their superior, monitoring by their superior, or a complete lack of motivation to work with the particular device. Thus, at the end of the day, the design of the product or space might be blamed whereas the reality is that the maximum output from the product could not be obtained because of the environmental factors mentioned above – all of which are intangible.

The physical environment comprises the tangible or measurable elements of the environment, like heat, light, humidity, ventilation, and so on, which affect manufacturing and also performance of products. If it is too hot then workers at the assembly line are unable to give their best. Similarly, if it is too hot inside a room where doctors are performing surgical procedures with their equipment then their efficiency may drop, not because of bad equipment design or poor skills but because of the environment. Another aspect of the environment is the context. A person trying to open a shampoo sachet with a wet hand will struggle to do so. This is because the manufacturers did not keep in mind the context in which the shampoo sachet would be opened.

The technological environment comprises that part of the environment which is consists of the different technologies that a human has to handle to perform his tasks, from heavy machinery, to small tools, to micro gadgets that amplify his capabilities or helps him to achieve a desired goal. In the technological environment the emphasis is more on the engineering aspects. This is finally followed by the psychosocial environment, which comprises the person's psychological state, such as mental well-being. A pilot under psychological stress inside the cockpit is more liable to make mistakes than another who is not, even if the equipment is well designed.

2.4 TYPES OF SYSTEMS

There are different books which discuss the various types of systems, but for the sake of ergonomics and design we shall consider some that are very pertinent to design (Figure 2.3). From this perspective there are three different types of systems: manual, mechanical, and automated. Within these categories there are further classifications of open loop, closed loop, etc. A manual system is where there is no automation at all. A farmer working in the field and harvesting the paddy with sickles best represents this. The focus of the designer should not only be on the design of the sickle but also the farmer's working posture, the duration of work, and the environment (extreme heat should not heat up the sickle handle). A mechanical system is where the machine is used to amplify human capability. This machine is a little-improvised version of the tool used in the previous system. A person using a lawn mower to trim the grass is an example of this type of system. The designer should focus exclusively on the design of the mower in terms of the handle height, safety, and other mechanisms with the context of use in mind; for example, if the lawn is wet. An automated system is where physical/manual work load is almost minimal and the user has to mainly monitor the system and intervene when things go wrong. A person in a power plant control room is an example of this type of system. He has to keep a vigil on the displays and intervene when things go wrong. There is minimal physical load but increased cognitive load on the users. An open-loop

FIGURE 2.3 Different types of systems.

system is where the loop of human–machine/product and environment is designed in such a way that any input given to the product/machine moves in a unidirectional manner and does not come back to the users. A catapult is an example of an open-loop system, where the particle once released does not come back to the user and is lost in the surroundings. A closed-loop system is where after input is given to the machine by the human, the machine also gives some output to the human, and this goes on in a loop. An Automated Teller Machine (ATM) for dispensing money is an example of such a system. Here, the user feeds in information through the keyboard and then the machine displays information relevant to the user's account on the screen.

There are certain ways to look at systems to understand their reliability. When the components of the system are in series (e.g., in a kitchen where the work flow happens in a sequence from one end to the other) the system is unreliable, although it is relatively cheaper (Figure 2.4). When the components are in parallel, then the system becomes much more reliable but relatively expensive. The enhanced reliability is because each unit works independently. Thus, failure of one part does not affect the functioning of the other parts. In system in series, the failure of one part leads to a cascading effect, which leads to the collapse of the whole system. Chain lights used in festivals are a good example of a system in series, where if one of the bulbs/light-emitting diodes fuses then the entire light chain does not glow. An example of a system in parallel is power backup at your home. In the event of a power

FIGURE 2.4 Systems in series and parallel.

failure from the electric supply company, one keeps backups in the form of invertors, mini-generator sets, or solar-powered lamps. All these components are capable of working on their own, but add to the total expenditure of the power budget of the household.

2.5 SYSTEMS CHARACTERISTICS

The famous philosopher H.L. Meckein believed that people think that for complex problems there are simple solutions, but this is not true. For complex problems one needs to go much deeper to address the different issues and thus solve them. Every system, and especially the human–machine–environment system that we are talking about, has unique characteristics. Unless these characteristics are understood well, problem solving at a systems level will be extremely difficult, and ultimately the design will fail to produce the desired effect. The following are some of the characteristics of a system from an ergonomics perspective.

We have seen that any system always has an aim or mission. We design products or services with this in mind. If one has to solve the problem of cleaning the floor of the house without having to stoop on the ground, then this becomes the aim of the system and it has to be analyzed in that direction. This type of approach is unique in ergonomics and design and is often known as the "outside-in" approach. This is in stark contrast to what one finds in engineering or technology, where most of the time the focus is too much on the artifact and less on the context and the target users.

Hierarchy is an important characteristic of a system. There are different levels in a system, and each level is associated with the others. In ergonomics and its application in design, this is very important. For example, if one looks at the family as a system then there are certain levels of hierarchy. The father is the main earning member and is the head of the family. Next comes the role of the mother, who makes a budget and takes care of every family member. The sons and daughters form the next layer, including elder and younger ones. It is normal practice that the elder siblings help the younger ones in studies and also give them company as and when required.

FIGURE 2.5 Context of use in a system.

The context of use (Figure 2.5) is the next important component of the system. Your mom has a tough time opening the lid of a jam jar at the breakfast table with wet hands. All of us are aware of how difficult it is to open a kitchen tap with oil, water, and soap on our hand – every time we try opening or closing it, the hand slips. Try operating your smartphone with thick woolen gloves on. Hence, these issues need to be considered at the design stage. The mental state of the person and/or the type of work wear they will be wearing while performing the set task with the product or in the space needs to be taken into account.

Each component that makes up a system serves a specific function. This is the beauty of systems ergonomics: although each component can function on its own, in unison their impact is multiplied. This is just like a soccer team. All 11 players play their role in the game, but that does not decide the fate of the team. It's only when all members play in perfect synchrony that victory comes to the team. So the performance by individual players is one characteristic of the system, and the next characteristic, that is, component interactions, is what one notices when a team plays well or badly, as explained before. In a soccer team it's very important that there is good coordination among all 11 players on the field. In the absence of such coordination it's not possible for the team to play well and win the game.

2.6 RELIABILITY OF THE HUMAN–MACHINE SYSTEM

In this entire systems perspective, the human is the weakest as well as the most important link at the human–machine interface, and has been found to reduce overall system reliability.

Thus, whenever there are humans in the system, the overall system reliability tends to drop (because of inherent human variation which exists between humans as well as for the same human every day). This makes the human the weakest link in the overall system loop of human–machine–environment. Designers need to take off from this point while applying principles of ergonomics in the design process. To do this, it is important to know the characteristics of the human body and mind when designing product and spaces, and that is what ergonomics is all about. The reason why humans reduce reliability at the human–product interface is the huge variation which exists between humans and also within the same human for performing day-to-day activities. For example, you have the same smartphone used by different types of people who vary in age, gender dimensions, intelligence level, and so on. Similarly, you do not have the same performance in a classroom. You cannot score the highest marks, always come first in sports, and so on. There are ups and downs in your life for different reasons. So, you are a different person every day.

2.7 THE PYRAMIDAL STRUCTURE OF THE HUMAN–MACHINE/PRODUCT–ENVIRONMENT SYSTEM

Now, if we look at the overall systems ergonomics from a design perspective, then we will see that it has a defined structure, in the form of a pyramid, having a base and an apex (Figure 2.6). The three corners of the triangle are made up of human, machine (product), and environment. A closer look at the figure reveals that the base of the pyramid is a little extended to form a rectangular base. This makes the system a little more complex than we talked about at the beginning. The user of a product or space lies at the sharp end of the pyramid (system) and at the base or blunt end of the system or pyramid is the designer or decision maker. This is the actual model of the system for a designer. So, whenever the user makes a mistake in using a product, then normally the onus is on the user and we call it "human error". The reality is that people at the blunt end of the system, that is the designers are also equally responsible for such errors made by the users. Therefore, the concept of systems ergonomics helps us to tackle such issues and enables the designers to come away from the blunt end and put themselves at the feet of the users, and then design their products. Here lies the justification of systems thinking in ergonomics. An example of this pyramidal structure of systems will make it clear. I was always fond of music and as I was born at the time when the Walkman was popular, I used it while jogging, and also when traveling. When I was traveling, I wanted to listen to music at night.

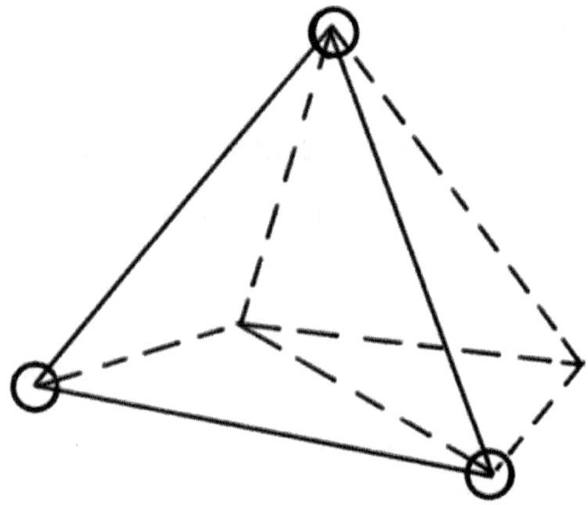

FIGURE 2.6 Pyramidal structure of systems ergonomics concept in design.

To my surprise I found that while lying down in the dark, it was almost impossible to locate the control buttons by just touching them. When trying to press play I was repeatedly pressing the eject button. This is a unique example where designers did not think about the users and the different contexts under which they would use the product. They remained at the blunt end and designed it from there, whereas the users at the sharp end struggled with the product. Thus, it was a systems-level problem and not the fault of the target user.

2.8 ASSIGNMENTS

1. Pick up any product of your choice and try to list how many different conditions the user needs to use it.
2. List some products around you which depict open- and closed-loop systems. Can you identify why they are so?
3. Draw a systems pyramid and indicate at which end designers lie and at which end the users of products lie in the case of your smartphone.
4 Identify the different subsystems that make up your family.

3 Human Body Dimensions

OVERVIEW

This chapter introduces the reader to the importance of human body dimensions in design. Techniques for using human dimensions for designing products are discussed. The concept of the percentile value in human dimensions as related to design is explained as well.

3.1 THE NEED FOR A HUMAN DIMENSION

In this world, everyone is different. This difference exists in terms of physical dimensions, such as height, weight, strength, and it can be in the form of differences in mental capabilities, and so on. There are differences in terms of knowledge, awareness level, and many things like that. The biggest challenge to the designer is to design product(s)/spaces in such a way that it caters to the needs of the entire spectrum of the population. It is not always possible to design for individual, and hence a prudent approach is required, wherein products/spaces are designed which cater to the needs of different people with different physical dimensions. Have you thought of what would happen if an elephant is made to enter the cockpit of an aircraft, a child wears his father's shoe, or a surgeon tries operating on a patient with a jumbo saw? These are all examples of a bad fit between human/animal body and the products. The biggest challenge for the designer today is to design products and spaces for a variety of different people.

3.2 ANTHROPOMETRY

Anthropometry is the scientific measurement (Figure 3.1) of different parts of the human body with specific scientific equipment. The term "scientific" here means that anthropometric dimensions are always taken with reference to certain "landmarks" in the body. These landmarks may be bony elevations or depressions in the body. So the specificity of this type of dimension is that it should be taken from bony elevation to elevation, from elevation to depression, or from depression to depression in the body. Different types of scientific equipment are used for this. The traditional types are known as Martins Anthropometric Kit, spreading calipers, sliding calipers, cones, special measuring, tape, and so forth. The latest ones are 3D scanners which enable one to take dimensions much more accurately and faster. Grid boards are also used for measurement. The subject stands against the grid board (which is a board with grids just like graph paper) and the number of grids (boxes) is counted to obtain the measurement.

3.3 PROCEDURE

There are certain set procedures for taking anthropometric dimensions of the human body. The subject being measured should be stripped to the waist and preferably

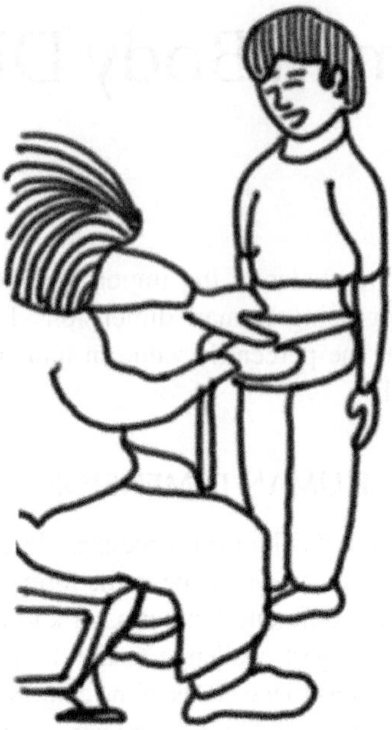

FIGURE 3.1 Anthropometric measurements.

wearing shorts. In the case of female subjects, a small garment covering the upper and lower parts of the body will suffice. It is absolutely necessary to take prior permission from the subjects and explain to them what is going to be done and how. In the case of female subjects it is absolutely necessary to have only female investigators for these measurements.

3.4 TYPES OF ANTHROPOMETRIC DIMENSIONS

There are two types of anthropometric dimensions, static and dynamic. Static, as the name suggests, is the dimension of the body taken with the body at rest without any movements. On the contrary, dynamic anthropometric dimensions of the body are taken with the body in motion, or more precisely considering the range of movement that the body will exhibit with the particular product or in the space. For example, if you are to decide upon the length of a door handle (Figure 3.2) then the palm breadth has to be considered. If you just take your palm breadth and fix the length of the door handle there could be problems. The reason behind this is that when a person grabs a handle with force then a portion of the tissue (flesh) of the palm protrudes on both sides, thus increasing the overall palm breadth, which needs to be accommodated within the length of the door handle. Thus, there is dynamicity in the movement of the hand. Similarly, when you are walking down a narrow aisle, you do not move in a straight line with your feet straight (it is difficult doing that!). There is always some lateral sway of the body, which needs to be factored in when designing the aisle width. For entering a car, no one can enter in perfect symmetry. Some people

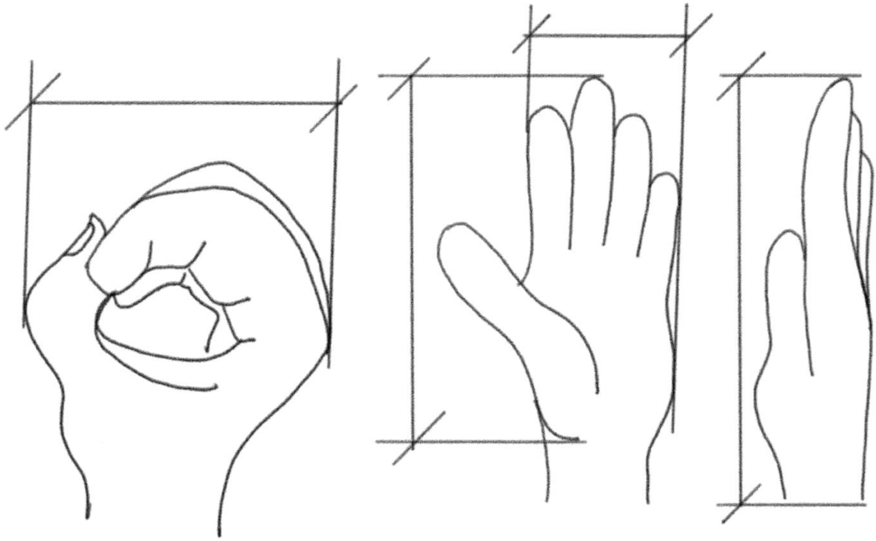

FIGURE 3.2 Anthropometric measurements for a door handle.

throw their buttocks in first, some put their head inside, and so on. In essence, all dimensions start from a static position (rest) and then move into dynamicity, which is the movement from rest to the point at which the body has moved further and finally rested. This is the most important part of anthropometry that the designer has to consider.

3.5 PERCENTILE VALUE

The biggest challenge to the designer is to design products and spaces for a wide variety of people, all of whom are different in terms of body dimensions. It is not always possible to customize for individuals, so generalization has to be done, but on what basis? In this world, statisticians have noticed certain trend in the population. If you plot frequency of occurrence (how many times) of certain human body dimension on the *y*-axis (vertically) and the human dimensions, for example, for hand length, on the *x*-axis (horizontally) then a bell-shaped curve results (Figure 3.3). This is called the normal distribution. The *y*-axis is called the dependent variable because the value of it depends on the value of the *x*-axis which is called the independent variable. Why is the value of *y* dependent? This is because the frequency (number of times) of occurrence of a particular anthropometric dimension is dependent on the type of anthropometric dimension in a given population. For example, the number of people having their body parts between 100 mm and 80 mm is dependent on the exact body part, and not the other way round. The bell-shaped curve is also called the normal distribution and is the normal characteristics of a given population when it comes to their body dimensions. If you look carefully, the curve is at the maximum at the center. This is where the majority of the values (body dimensions) hover. The exact middle point is the mean (the central point of the data, or the average as we call

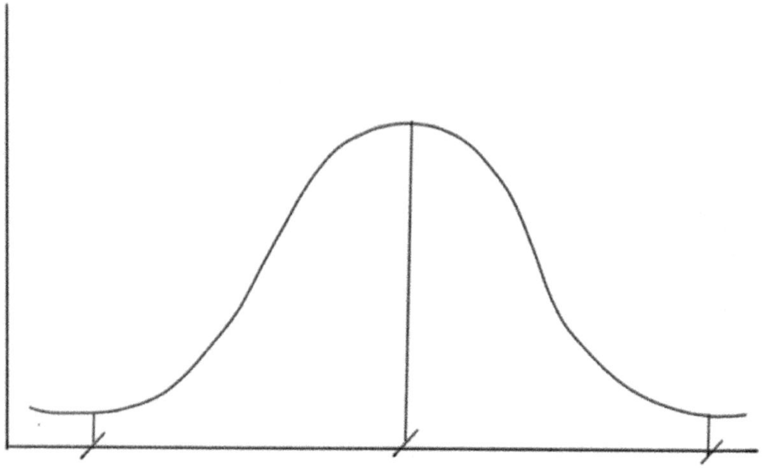

FIGURE 3.3 Normal distribution curve.

it). The curve tapers at the two ends of the mean. In essence, the area covered by the curve represents the population and the probability (chance factor) of them having different anthropometric dimensions. At the central part, the chances of having the majority of the population with similar dimensions are higher compared to the ends. On the left side the values decrease, and on the right side they increase. This is the technical detail of the percentile value. So, this curve tells us how the values are spread out.

3.6 THE NEED FOR PERCENTILE VALUES IN ANTHROPOMETRY

You are very hungry and desperately looking for food in the middle of a jungle. Unfortunately, there are no restaurants in the vicinity. You manage to get some vegetables in the forest, but you cannot eat them raw. This is at times the condition of the designer. They have a lump of data of different anthropometric dimensions but do not know how to use them for designing a product or space (the situation is like your hunger in a jungle). The raw vegetables represent the lump of data from the population, which you have with you, it is valuable like gold but not in a ready-to-use format. These vegetables need to be cooked to aid in easy digestion. The next stage is cooking the vegetables with oil and spices. This is similar to the statistical treatment of the lump of data to check for its reliability. Why reliability? So that the data truly represents the entire spectrum of the population for whom we are going to design. This step is very important, because human body dimensions are taken from only a few people. We cannot go to each and every person to measure their body parts. It's a risk, taking the dimensions of a few people from the population when we are designing for the entire population. This is where statistics comes to our rescue. Statistics is cooking the vegetables. The food is still not ready to eat until it is served on a platter. This is the statistically treated data, which confirms the data reliability and further confirms whether it truly represents the entire spectrum of the population

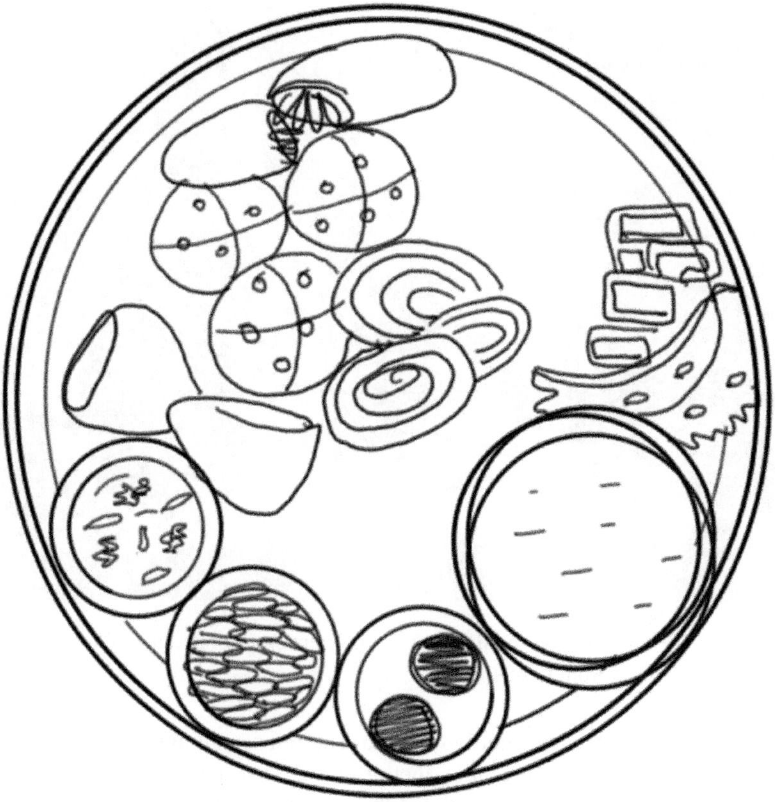

FIGURE 3.4 Making the raw anthropometric data useful for designers.

or not. The food served on the platter can now be readily eaten (Figure 3.4). This is similar to when the statistically treated anthropometric dimensions are plotted on a scale of 0 to 100 for ease of use in design, which is called the percentile value (the word comes from a scale of 100). In the absence of a percentile value, designing for the mass becomes difficult, as designers get confused when and in what manner to use the data. Suppose you want to design a chair for the common man. For that you need to have to hand the maximum and minimum anthropometric dimensions of the different body parts pertinent in design. The challenge is that everyone, starting from the fat to the thin, the tall to the short, male and female, should be in a position to use the chair with ease.

3.7 THE IMPLICATION OF PERCENTILE VALUES

Percentile values are important in designing products/spaces for the masses, but are not as straightforward as many people think (Figure 3.5). There are people of different body dimensions in a store, but the dimension of the store has to cater to the needs of all types of people. The first mistake a student of design makes while using the percentile

FIGURE 3.5 Deciding the percentile value for different types of designs. (Photo by Lukasz Dziegel from Pexels.)

value is to pick up all 50th percentile values and design accordingly. The logic is, if all dimensions are in the 50th percentile, then it will cater to the majority of the people (lying close to the mean in the normal distribution curve). This type of decision is disastrous for design. The human body is made up of body parts which are not in exact proportion to one another. In fact, the proportion of different body parts for different people also differs. Therefore, any individual is a mixture of different percentile values when it comes to anthropometric dimensions of body parts. For example, your height might be in the 95th percentile, palm length 50th percentile, shoulder breadth 75th percentile. This is reality, you are not symmetrical! Thus when you design you have to keep this in mind. A person of stature of the 50th percentile is not guaranteed to have all their anthropometric dimensions in the 50th percentile as well. Therefore, the target population for whom you design has body parts of different percentile values! That means an individual can have body parts of different percentile values!

3.8 WHEN TO OPT FOR WHICH PERCENTILE VALUE IN DESIGN

Opting for the correct percentile value is a big challenge for the designer. For the sake of simplicity we will now refer to higher percentile and lower percentile anthropometric data. Any percentile value above the 50th percentile will be referred to as the higher percentile value and any value below it will be referred to as the lower. If you were to design a screwdriver handle, the length of the handle will be guided by the palm breadth and the cross section of the handle will be guided by the inner diameter of the grip. Let's take the case of handle length. If palm breadth is our reference dimension, what happens if the handle length is too small? Only a small hand (palm breadth, to be specific) can hold it with ease and for someone with a larger palm breadth it will dig into the palm. Both conditions are bad, but say that you have to choose between these two "worst-case" scenarios, which one would you go for? To come to a conclusion, let us look at the two cases again. If the handle length is too small, it digs into the hand for the larger hand. If the handle is too long, the small hand can still hold it, although with difficulty. Therefore in this case, we opt for the higher percentile value (we do not fix at this, but take this as a reference point and then optimize later). Now let us move onto the diameter or cross section of the handle. If the handle is too thick, the small hand cannot hold it; if it is too thin, the larger hand can still hold it with difficulty. Thus, we opt for the lower percentile value. So, the rule of thumb is when it has to do with access or reach we opt for the lower value, and when it comes to clearance we opt for the higher value.

Let me give you an example from space. You are to design a public telephone booth. For this purpose you have to decide upon the door width of the booth. You consider (we will talk more about these later!) span akimbo length (distance between two elbows when extended sideways). In this case, if a person with larger span akimbo length can enter the booth, then the one with a smaller dimension will have no problem. Therefore, we opt for the higher percentile value. Now, you need to decide upon the height of the latch for locking the door from the inside. Here you need to consider the reachability of the shorter person, and then the taller one will have no problem. I would again like to emphasize that we are talking about opting for a particular percentile value and not "fixing" it there for final design. There are many more steps toward that, which we are going to discuss now.

3.9 STEPS IN CALCULATING THE PERCENTILE VALUE FOR DESIGNING PRODUCTS AND SPACES

Imagine that we have the database of anthropometric dimensions at hand. We now have to calculate the correct value for our product/space and design it accordingly. There are many stages and the following is a rough guideline toward that. So whether you are designing a new product/space or redesigning the same, the process will act as a guideline for you (Figure 3.6).

Step 1: Let us take the example of a door handle. You have to start with a "task analysis", that is, observe how many different ways people operate a door handle and under what different contexts. You might notice in this step that people hold the handle in different ways, and with different parts of the palm, and how the handle height affects the body posture. This is when you have a reference handle. What if your product is new, such as a door latch to be operated by the foot? In that case you have to decide upon the touch point between different parts of the feet and the product and then note down the relevant anthropometric dimensions which would be relevant in designing the product. In the latter case it would be an process involving much iteration, whereas in the former the method will be the same with less iteration. This stage helps you to identify the relevant anthropometric dimensions in that particular design.

Let us now take the example of designing (not redesigning) a study room for yourself. Now you have to consider the movement pattern of the people in the space,

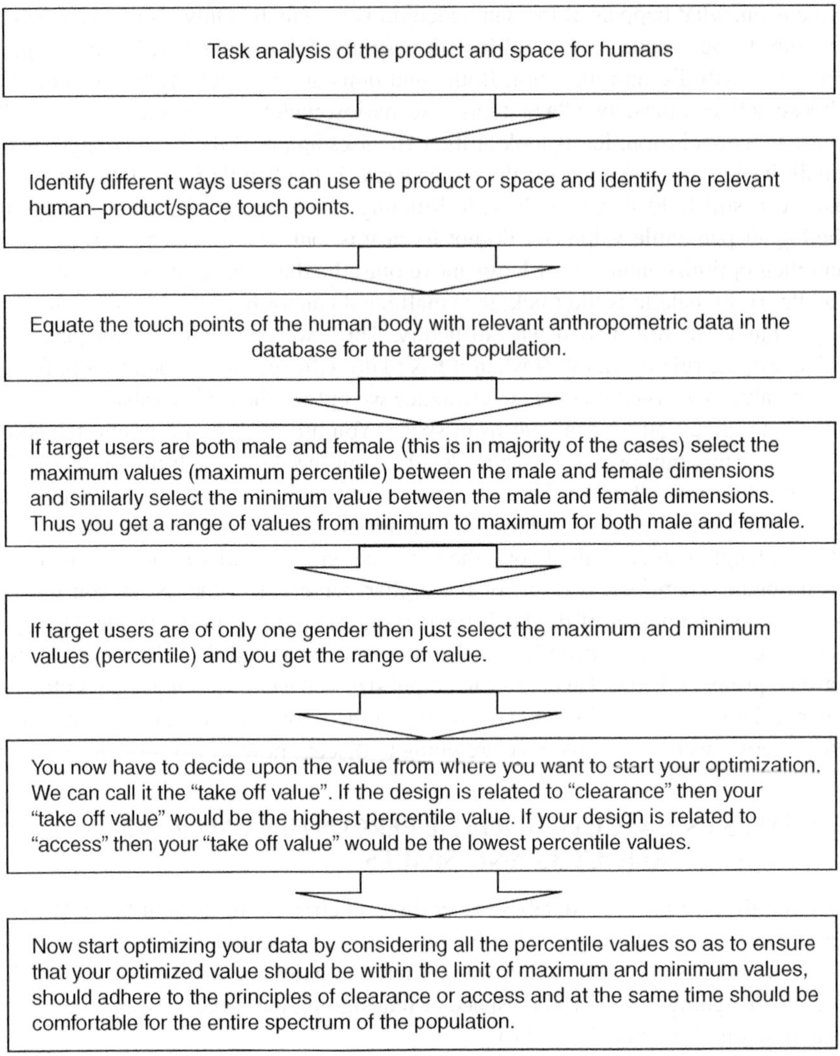

FIGURE 3.6 The anthropometric design process – a schematic representation.

different types of body postures including extreme ones so as to get an insight into the allowances (dynamicity), and also identify the different touch points or relevant anthropometric dimensions. If a reference space is available it is much more similar – you have to observe peoples' movement in that space. If a reference space is not available, you have to "imagine" or draw mannequins (small-scaled models of humans) and move them in your space to identify the above points.

Step 2: You now identify different touch points, draw them, and study them very closely. Then you try to establish a link between these touch points in the body and the closest possible anthropometric dimension in the database. For example, if you are to list the relevant anthropometric dimensions for designing a hammer handle then the probable anthropometric dimensions relevant to its design would be palm length for deciding the length; grip of the inner diameter for deciding the cross section of the handle, and so on. For a space such as a telephone booth the dimensions could be stature for the height of the booth; span akimbo for the door width of the booth, and so forth.

Step 3: Once the anthropometric dimensions are identified, go to the relevant table of data for each dimensions. A close look at the table will reveal that there are five different percentile values given for the male and female populations in any context. So there are male values, female values, and combined male and female values. Write down the table of data for each dimension.

Step 4: Ask yourself who your users are: Male? Female? Both? If there are only male users, things are relatively simpler than if there are both male and female users. Let's start with the complex case first and imagine that our target users are both male and female. You are now ready with the relevant anthropometric dimensions of each body part and their percentile values. Now comes the issue of value selection and optimization.

Step 5: You learned the two rules in percentile value selection: for reach, lower percentile value; for clearance, higher percentile value. These are the "take-off" points for percentile calculation and we will never fix the dimension at this. For the lowest (e.g., 5th) percentile value for both male and female, look for the lowest value. If the female value is lower, circle it: this becomes your lower percentile take off value. Again, for the maximum (e.g., 95th) percentile select the higher value among the male and female 95th percentile. Circle the male value, if it is higher. Now you have a range of percentile values, female lowest, male highest. You now have to optimize your dimension within this range.

Step 6: Look at the intermediate value for both males and females. If it is a question of access, try to optimize it a little toward the lower percentile, and in the case of clearance try optimizing it a little toward the higher percentile value. Do not forget to add an allowance for dynamicity of movement, clothing, and gloves. This will depend on your user study and the task analysis done in step 1. Remember to fix your dimension to the nearest whole number for manufacturing feasibility. So if your anthropometric dimension turns out to be 112.2 mm then round it off to either 110 mm or 120 mm depending upon your observation and task analysis and of course the manufacturing feasibility of the product or space.

3.10 SOME MORE

If you have gone through the anthropometric dimensions seriously, you would have noticed that there are "gaps" in between percentile values. For example, in many contexts there are percentile values, 5th, 25th, 50th, 75th, and 95th. So what happened to 6th, 7th, 8th … 26th, 27th … 51st, 52nd values? This pattern might be a little different for other population groups. The basic reason for this is that from the 5th percentile (in this example) to the 25th percentile (in this example) there is very little change in dimension (not significant). Hence, these are not relevant for the designers and have not been given in the data table. The main aim in anthropometry is designing for the maximum spectrum of the population. If you remember the bell-shaped curve of the population, then we will see that in this

FIGURE 3.7 Different somatotypes of the human body.

design process some people will be comfortable with the design, but those at the tail ends (extreme short or tall) will be a little uncomfortable. This is where you have to make a decision depending on the product you are designing. In exceptional circumstances when it comes to customizing the design, you need to design for the extremes of the population, such as the smallest and the largest for the respective anthropometric dimensions. Examples include emergency exits, gates or doors in public places where it is possible that extremes of the population would also come. In an emergency, every person should be able to move out of the space. The best design is providing adjustability into it so that each person can customize the design as per their needs. This is not possible in all cases. In such cases, the second option could be the one discussed above.

3.11 BODY SOMATOTYPES

We have missed out an important topic that talks about categorizing human body based on the amount of muscle, fat, and bony material present (Figure 3.7). This type of classification is called somatotyping. There are three major types of body somatotypes: ectomorph, mesomorph, and endomorph. An ectomorph is a very skinny person. An endomorph is a bulky person with a lot of fat deposits in the body. A mesomorph is a muscular person. These somatotypes need to be considered when designing product and spaces to ensure that there is enough space for people to navigate and account for dynamicity in body movements.

3.12 APPLICATION

By this time you might have realized the importance and application of anthropometry in design. It is a complex subject and we have tried to simplify it for your profession.

There are other elements of percentile calculation, mannequin construction, and 3D modeling which we have deliberately left out. We will use more principles of anthropometry in the relevant sections.

3.13 ASSIGNMENTS

1. Take a hammer and find out the level of anthropometric mismatch with reference to the dimensions in the database.
2. Select a small space, such as your bathroom at home and perform an anthropometric analysis of the same.
3. Optimize the height of a dining table at home based on the different types of users.
4. Optimize the height of a grab rail of a public bus on which you commute regularly.

4 Ergonomic Principles of Hand-Held Products

OVERVIEW

This chapter gives an insight into the different ergonomic principles that guide the designer in designing hand-held tools and products. The different design and non-design aspects of ergonomics are also explained in this chapter.

4.1 ERGONOMICS AND THE HAND

The human hand is one of the most developed parts of the body after the brain. Primitive man used the hands extensively for hunting and gathering food. Later, hands had to do perform a different set of tasks with lots of forceful exertions and repetitive movements. The structure of the hand is not meant for activities such as repetitive movements and forceful exertions. This leads to a mismatch between product and hand, leading to different types of ailments which began manifesting themselves as pain, numbness, fatigue, and ultimately affected human performance and well-being with the product.

A close look at the structure of the human hand will make things clear. The hand has four different parts: shoulder, upper arm, forearm, palm, and the fingers. These are all made of bones which are supporting structures. The wrist (Figure 4.1) is made up of a number of bones called the carpels. Immediately after this are the metacarpals, which extend into the phalanges or fingers. The bones give support to the hand and there are muscles in the upper arm and forearm which helps in movement of the bones at the joint. The important point to note from the design point of view is the carpal bones, which creates a tunnel or bottleneck between the palm and the forearm. All movement of the hand is coordinated by the brain through nerve connections. We now discuss the different ergonomic design issues involved in designing a hand-held product or tool. It can range from a simple spoon to a gear control operated by hand. The basic ergonomic issues are the same, with changes at places which will be discussed in the relevant sections. A word of caution is that these are all gen eral principles, and should be applied with a lot of care and, most importantly, they should not be applied at random in all designs.

4.2 SOME ERGONOMIC PRINCIPLES

1. Relevant anthropometric dimensions. The relevant anthropometric dimensions of the hand should be taken into account in the manner discussed in the previous chapter. The degree of dynamicity should be accounted for. This should be done at the prototyping stage and throughout the design process. For example, if you are designing

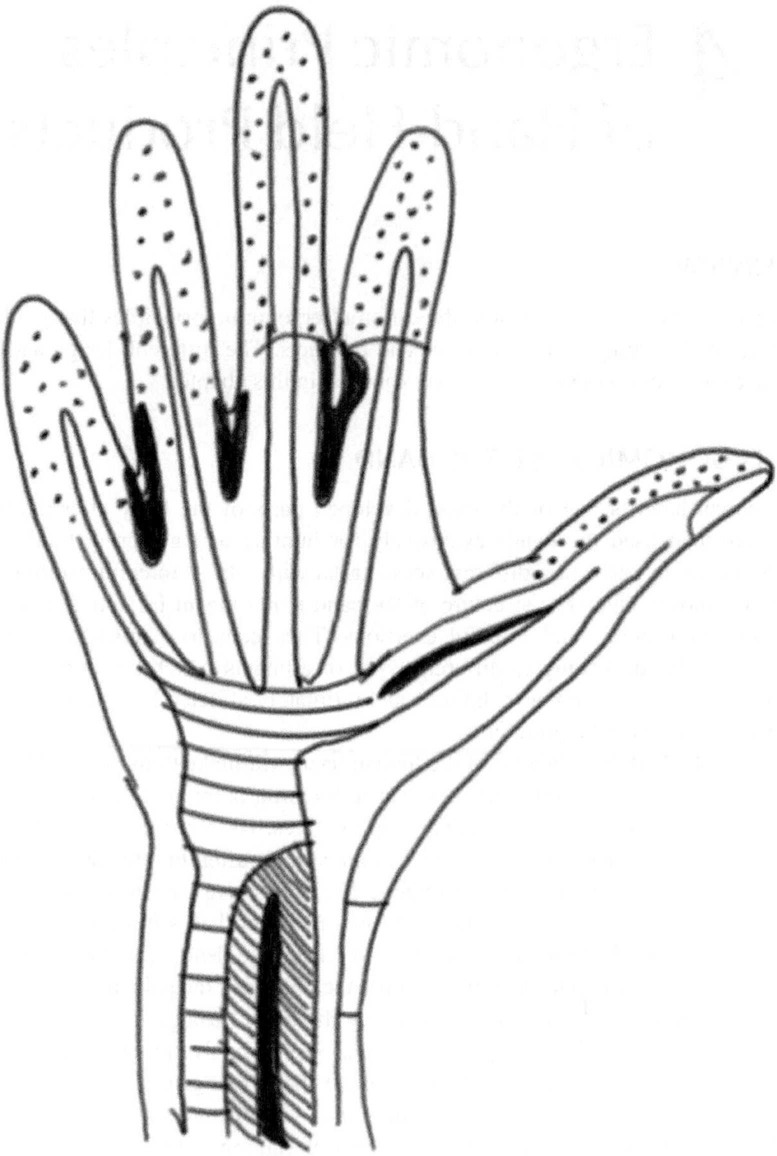

FIGURE 4.1 The human hand.

a coffee mug, then you need to take into account the anthropometric dimensions of the fingers and also the grip inner diameter to design the handle diameter as well as the clearance of the fingers from the main body of the mug. The shape of the product should match the shape of the hand, especially the palm. If you are designing the handle for an umbrella then the handle would bulge a little at the center and taper toward the two ends. This ensures a proper grip with the palm of the hand, which is a little concave at the center.

FIGURE 4.2 The neutral wrist position. (Photo by Roman Koval from Pexels.)

2. *Proper wrist position.* Any product that needs to be held in the hand should be designed in such a way so as to keep the wrist at a neutral position (if you remember the bottleneck at the carpal tunnel I just mentioned, this is to prevent the bottleneck). The neutral wrist position is roughly at an angle of 70° with the horizontal as shown in Figure 4.2. How do you gauge whether your wrist is at neutral? There are two ways of doing it. First, give your friend a pen and ask him to hold it while sitting down at a table and resting his forearm on the table surface. Then ask him to keep his hand in the same position and let the pen drop on the table itself. The angle sustained by the palm on the table is roughly the neutral wrist position of 70°. For the second example, ask your friend to stand up and let the hand fall freely by the side of his body. The position now taken by the wrist and the palm is roughly the neutral wrist position. That is the reason why gun handles are a little tilted. When a hand-held product is designed, you need to keep this in mind and design the product accordingly.

3. *Static and dynamic muscle load.* The muscle of the arm helps with movement of the hand. It also exhibits a property when it is moving and this is called the "milking action". In this, the muscle alternately contracts and relaxes and this helps in proper flow of blood and thus removal of toxic elements, so the onset of fatigue is delayed. On the other hand, when the hand is not moving or remains unsupported, then the

FIGURE 4.3 Static and dynamic muscle load. ((a) Photo by Craig Adderley from Pexels. (b) Photo by Oleksandr Pidvalnyi from Pexels.)

same muscles remain in a state of continuous contracture (Figure 4.3). Blood flow is slowed down, metabolites start accumulating very quickly, and the hand is fatigued. I provide a small example now. Ask your friend to extend his arms forward. Then put two bricks in his two hands and instruct him to remain standing in that position. After some time you will notice that his hands start trembling, and finally he is unable to hold the bricks any longer. This is an example of static muscular load on the arm muscles. If you provide support from below through two bamboo poles you will notice that he is now able to hold the load for a longer period of time. Hence, support to the arm eliminates static load and is beneficial. Even if you are to extend your hand as mentioned in the example above, then it will also become difficult to sustain this position for long. To keep the limbs in an overhang position (without any support) the muscle has to remain in a continuous state of contracture. If it relaxes a little the position cannot be sustained. This is why static load increases in this posture of the arm. Therefore, when designing products you have to be careful to eliminate static load as much as possible. For example, if you are designing a garden pruner, then make sure that the user works with the pruner with hands close to his body. If he has to cut plants away from his body, provide the pruner with longer handles so that the hands remain close to the body. Weight reduction of the product can also help in reducing static load. For example, if you are designing an electric razor, then a heavy razor will cause fatigue much earlier than a light one. Have you ever tried using a vacuum cleaner at home? When you clean the floor it is not much of a problem, but the moment you start cleaning the ceiling of the room your shoulder and forearm start aching. This is because of the unsupported arm leading to static load on the muscles, while cleaning the ceiling.

4. Contact surface. If you play cricket, you will have noticed how good it feels to play with a bat which has textures on it compared to the one which is plain. If you have ever used a screwdriver with a textured handle, you might have noticed that compared to a plain handle, you are in a position to apply greater torque with the textured handle. Why are textured designs better? The reason is that we have certain receptors (special organs all over the body and especially in the hand) called tactile (touch) receptors. The function of these receptors is to receive information from the outside world, convey it to the brain, which after synthesizing the information instructs the body to act accordingly. In this example it instructs the hand to grip or exert torque according to the task and modulate the force on it accordingly.

5. *Position of the body.* The bones and muscle of the body are all interconnected. Thus, movement of one set of limbs affects the movement of the other set. For example, you are able to exert much more push force in some cases when you stand compared to when you sit. When you change your position (called posture) from seating to standing, or in any position, it affects the amount of force you can exert. This should be kept in mind when designing products. For example, a refrigerator door will be opened (unless the person is wheelchair-bound) while standing. Hence, at the design stage you have to account for the fact that the amount of force (minimal though it is in this case) will be while standing.

6. *Force requirement.* Have you noticed that your mom has a very tough time opening a new jam jar and your dad has to come to the rescue most of the time? If you analyze this you will see that in opening the lid of the jam jar, a grip and torque (anticlockwise direction) is required. Your mom's forearm muscles are structurally weaker than your dad's. This suggests that like human body dimensions, the force exertion capacity of humans also varies from one male to female which needs to be considered at the design stage. Hence, at the design stage you have to incorporate the fact that the amount of force has to be in tandem with the population percentile force data available in the market, with reference to males and females. If its not, then you have to generate one yourself by conducting a series of experiments.

7. *Work surface/orientation.* The orientation of the work surface will change your posture and hence affect performance. You use a marker pen to write on the whiteboard. Hence, the design of the pen should account for the fact that the surface is vertical and not horizontal. The grip position and thickness have to be in tandem with this orientation of the board. For example, while writing on the board, people seldom rest their wrist on the board surface, which is in stark contrast to what one does while writing on a piece of paper on a horizontal plane.

8. *Message conveyed by the form of the product.* I have often found "please push" written on the handle of the door of many offices. It sounds funny to me. If you want people to push a door to open it, why not remove the handle? Then there would be no other option but to push the door (you cannot pull a door without a handle!). If you design a switch which is concave then it automatically intimates to you that it has to be pressed with the finger to operate. This way of designing products so that the product "speaks" how it should be to operated is very important in design and makes it user friendly.

9. *Color and culture.* A product is never complete without proper colors. Colors add life to any product. When a particular brand of pressure cooker was launched in a country it was colored dark gray (close to black), though it was a beautiful design (in terms of its shape and utility). Unfortunately, this product did not do well in terms of sales in the country. When investigated, it was found that the maximum sale of

pressure cookers happens in that country during the wedding season, during which it is gifted to the newlywed couples who are about to start their family life together. The color black (or dark gray) is an inauspicious color in that country, and as marriage is an auspicious event, no one liked to gift this product. This is not because its design was not good, but because of its color. I remember a similar example, this time related to motorbikes. In rural India, it is a custom to gift motorbikes to the would-be groom. Initially, these motorbikes were all black in color. Amazingly, when the color of these bikes was changed to red, blue, green, and so on, sales picked up tremendously. The reason here was also cultural, that is, black was considered by many as an "evil" color, and hence many people were not willing to use it as a gift for the wedding. These examples show that you have to be careful when selecting colors for your product so that they are along the lines of the likes/dislikes of your target users.

10. Safety. Do you know the most "misused tool" on the globe? It's the screwdriver. Amazed? Yes, apart from screwing and unscrewing, people use it for making holes, cutting small products, digging nails, and scores of similar activities. This at times renders the product unsafe, whereas in reality it might not be. If you use the screwdriver (or to be specific, misuse!) for removing your ingrown toenails and then sustain an injury, it is not the fault of the product. In an attempt to make a product safe, screwdrivers are now available in the market with defined shape, which can be used only for screwing and unscrewing purposes and not for anything else. Many might argue that this type of use of a product only speaks about its multiple usages and hence is good, so what is the problem? There are problems. You have to understand the difference between "misuse" and "multiple use". Misuse is when a product designed for one specific purpose is used for other purposes for which it has not been designed. In an attempt to do this, the main functionality of the product is lost. For example, if you use the screwdriver for making holes in cans, then ultimately the sharp end of the screwdriver will become blunt and it won't be in a good position to fit onto the head of screws. On the other hand, if you want to design a screwdriver for multiple uses then possibly you will have to design some special attachments along with the tools for serving those functions.

11. Grips. There are different ways people can hold a product. It all depends on the intended purpose of the product. If you are to design the handle of your bike then you need to grip it firmly and thus the fingers are curled around the handle and touch the base of the palm. This is called a "power" grip, as the name suggests, when you require power or need to hold on to something with lot of force. There is a pinch grip or precision grip, when fingers come close to one another for doing precision work. You notice pinch grip when you paint a picture and all the fingers come close together. You will notice the power grip when you hold a hammer, with all four fingers curling around the handle and touching the thumb. Thus, the rule of thumb is when you design for power, go for power grip, and when you design for precision, go for pinch grip.

12. Gender. You know that structurally (most) males are stronger than (most) females, due to many factors, one of which being the presence of a larger quantity of muscle mass. Because men are stronger than women then any device which requires application of any type of force (grip, torque, push, pull, etc.) should take into account the maximum and minimum force-exerting capabilities of the male and female populations and design accordingly. The example of your mother trying to open the jam jar which I cited before is pertinent in this case. Therefore, next time you design a product, make sure you know your target user; if they are both female and male, then you have to calculate the amount of resistance in your product based on the available strength data of the male and female populations.

13. Receptors. The name suggests something which receives. Receptors are tiny organs in the human body that receive information from the outside world and relay it to the brain. The information is then synthesized and the body reacts. There are different types of receptors and of these the tactile or touch receptors are pertinent in product design. We have already seen the example of a textured bat handle and how it feels good. Previously, gloves were designed to protect workers from the high heat of tools and equipment in front of the blast furnace. As the gloves were thick, movement of the palm was restricted and they could not hold the equipment with ease. If they worked with bare hands then they would have to bear the intense heat. So, special gloves were designed where the portion in between the fingers was thin, thus facilitating coordination between the tactile receptors located in between the fingers and thus facilitating movement of the palm.

14. Context of use. Have you tried opening a small shampoo sachet while taking a bath? You would have had a harrowing experience. Ultimately, many of you open it with your teeth! This is because it is difficult to open the sachet with wet and oily hands when taking a bath. The designers never thought that shampoo has to be applied after you wet your hair and not before. The context in which your product will be used has to be factored in during the design process.

15. Actual usage of product. In some countries where there is a lot of dust, people have the habit of covering every precious thing. Smartphones are put in a case (they take them out every time they use them), the television is covered, and the microwave oven is covered. The remote control for the television is also put inside a cover. The reason for this is protection against environmental agents, namely dirt and dust. Try operating the television remote with the cover on! You will have a tough time pressing the buttons, as you do not get adequate tactile feedback. At the design stage you have to think of this, and even think of designing the cover, which is as important as the device itself.

4.3 ASSIGNMENTS

1. Select any hand-held product such as a paper cutter, scissors, or hammer and list the good and bad ergonomic issues in them.
2. Develop a list of ergonomic issues that one needs to check for in a hand-held product like kitchen tongs.
3. Select a tap to be used in the kitchen and analyze its ergonomic issues from the viewpoint of context of use.
4. Analyze a nail cutter for your toes from an ergonomic viewpoint and suggest some ergonomic changes for it.

5 Ergonomics of Space

OVERVIEW

This chapter provides an overview of different ergonomic issues in designing any human-made space ranging from a bathroom to a restaurant. There is a specific focus on the ergonomics of seating as well as different behavioral space dimensional factors.

5.1 HUMANS IN THE CONTEXT OF SPACE

Space is defined as the area around where a human lives, works, or performs some activities. Space may range from local (e.g., the doctor's office), intermediate (e.g., a big university), or it may be general (e.g., a village or town). So, when a human lives in the space, the space needs to be designed keeping in mind some ergonomic principles. Have you noticed how customers are made to sit at a hair salon? The chairs are placed quite far apart, so that the barbers' elbows do not touch each other, as this might affect the quality of the hair cutting. Therefore, when designing any space, from a small space inside a car to a large space inside an auditorium, certain ergonomic principles have to be adhered to. Each person needs a specified area to work or to live called a human work-space envelope or personal space. Hence, it's prudent to keep different elements depending on their importance of use with reference to this space. Some ergonomic principles related to design of space are briefly discussed below.

1. Visualizing the space. Any space has to be perceived with reference to the human body in two planes. From the top, the planes are also known as plan view, and from the side, back or front they are known as elevation (Figure 5.1). The plan view gives you an insight into the reachability aspects of the human body in the seating or standing position (also called standing and seating reference points, respectively). The elevation will provide an insight into the height of different body landmarks (called anatomical landmarks) with respect to the different landmarks in space (discussed in Chapter 3) and the elements in space (chair, table, cabinet, sofa set, etc.). For example, let us say that you are going to design the kitchen in your home. First you have to visualize the space. You start observing the space from the top with your mother cooking. Now in that view, you see her head and her arms stretching out for different utensils. From this you get an estimate of how to place the utensils within her reach, which utensils can be kept out of reach, and so on. The side or frontal elevation will tell you how high or low the kitchen platform is for your mother. Is the rack for keeping the spices kept at a very high point? This stage prepares the ground for further analysis.

2. Anatomical landmarks. There are three main anatomical landmarks with reference to which any work-space height can be fixed (Figure 5.2). Before fixing the work surface height you also need to consider the type of task that is to be performed

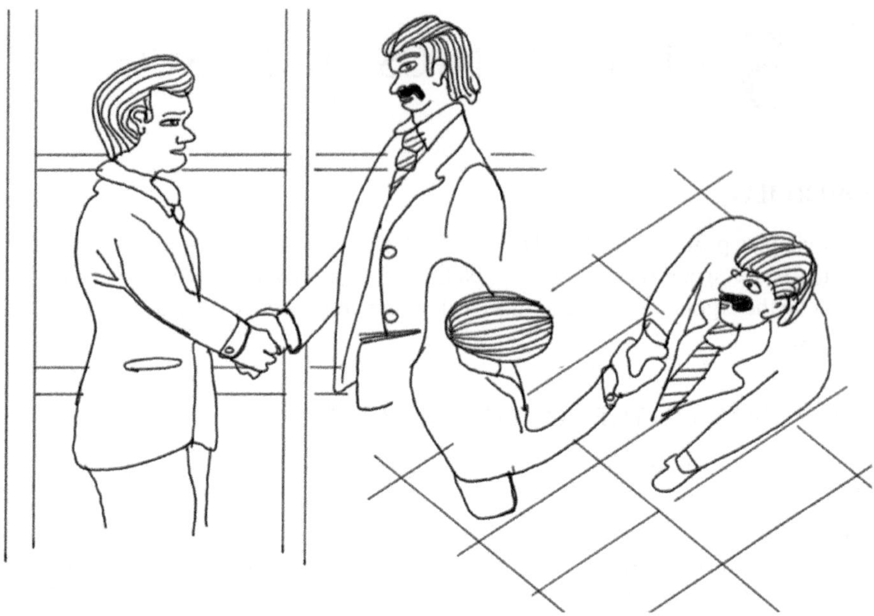

FIGURE 5.1 Space visualization for a layman.

at the work surface. The three anatomical landmarks from the lower part of your body (feet) are trochanteric height (the place near your hip where you wear your trouser), elbow height (height of your elbow), and substernal height (the lowermost depression between your two rib cages; where you experience pain after indigestion!), all measured from the ground. If you are performing some very precise work, for example, observing the controls of microwave oven in the kitchen, then the work surface height should be up near the substernal height toward the eyes. If the task is light, for example, using your laptop, then the work surface height should be fixed near the elbow height. If the task involves heavy work, such as using a planer on a wooden plank, then the work surface should be fixed near the trochanteric height. Now, as mentioned, you have to do many calculations in terms of percentile data, fixing the dimensions for work surface height at any landmarks. Landmarks are only reference points for you to use to design your space or the elements in space.

3. Access/reach in the work space. Every human has an envelope around them in which they carry out their tasks in space. This envelope is invisible, but the designer has to perceive it in order to design the space (Figure 5.3). In plan view, there are three work zones that have to be calculated from the seat or standing position as the reference point. These zones are primary, secondary, and tertiary. They can be roughly estimated by: keeping your elbows close to your body, the area that the forearm can cover is your primary zone; now extend your arm – the area your entire arm can cover is the secondary zone; and anything beyond this is the tertiary zone, which is any area which has to be accessed by bending or stretching the body. This is a very rough calculation for different work zones with respect to the human body. When

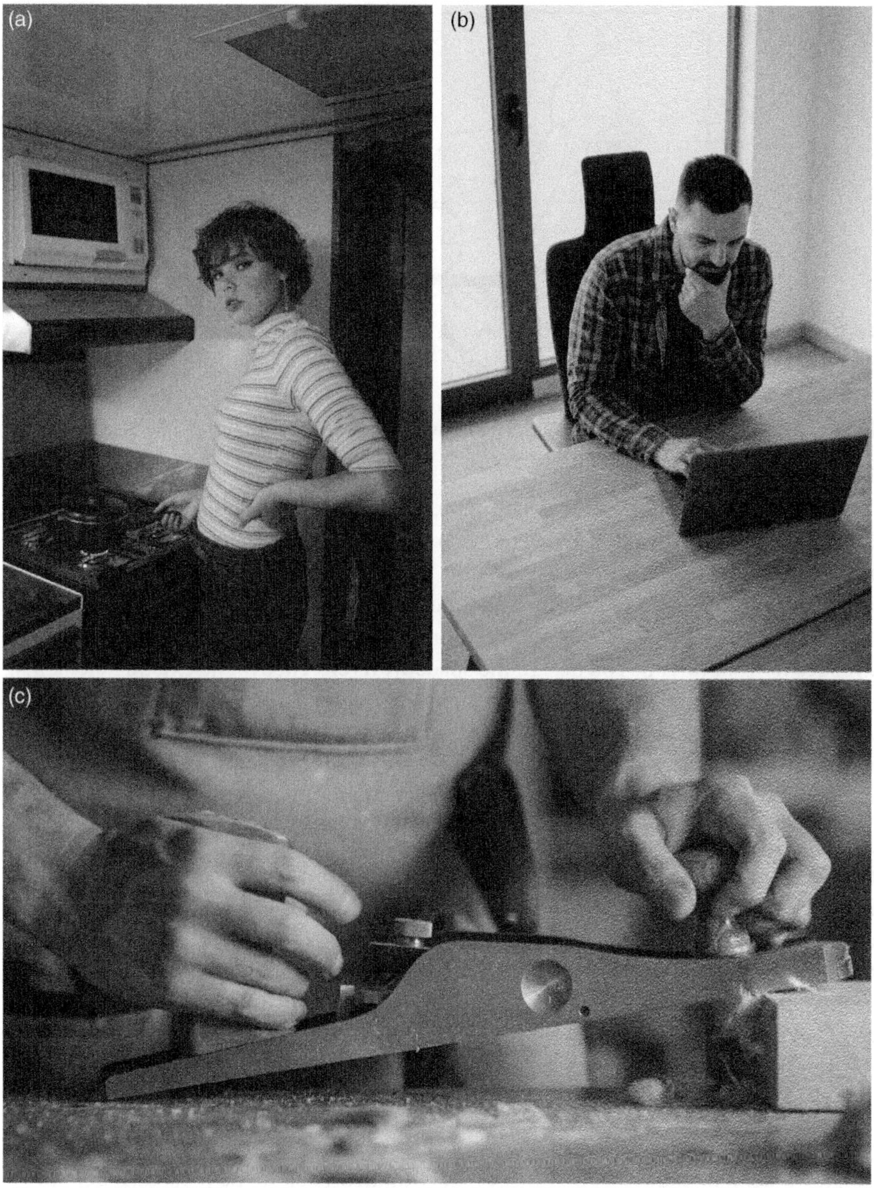

FIGURE 5.2 Anatomical landmarks for calculating work surface height. [(a) Photo by Pedro Sandrini from Pexels. (b) Photo by Djordje Petrovic from Pexels. (c) Photo by Burst from Pexels.]

designing your space the important elements have to be kept in the primary zone, so that you can access them with ease. For example, the keyboard for your computer is kept in the primary zone. In the secondary zone will lie the desk calendar, which can be accessed by stretching out your hand. In the tertiary zone you might keep your

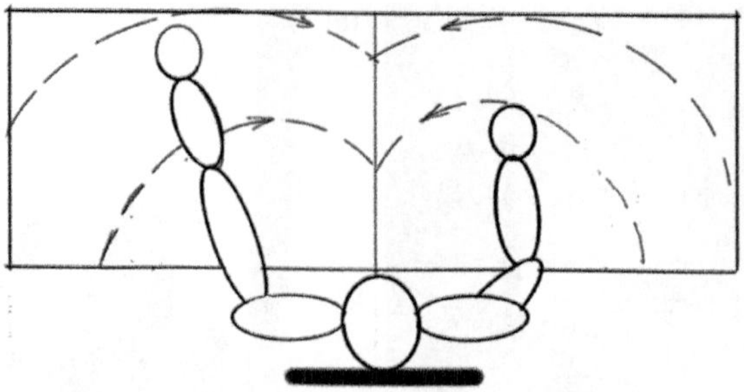

FIGURE 5.3 Anatomical landmarks for work space reachability issues.

books and files which you need to access every now and then and if necessary can be accessed by leaning or moving forward.

4. Dimensions and design of accessories. Any space is rarely empty apart from humans. There are furniture pieces such as tables, chairs, cupboards, beds, and benches. These furniture pieces are also called "elements" in space. As a space designer you also need to keep in mind the dimension of the elements in space. For example, if you are to decide upon the dimensions of a space, then there are two avenues in front of you. You design the space with reference to the existing elements, so that the elements fit in with ease and there is enough circulation space for users. Another instance could be that you have to design the space as well as the elements in space. In that case, it is first important to design the elements based on anthropometric dimensions and then calculate the space requirement.

5. Design of space within and without constraints. As a space designer you face two challenges. At times, you would be asked to optimize the space and suggest the number of people and elements that the space could hold, but without any increment of the existing space. Another instance could be that you would be asked to optimize the space with the given number of elements wherein you have to suggest the maximum number of people that could fit into the space. A third option could be that you have to optimize the space based on a specified number of people and also suggest the number and dimensions of the elements in space.

6. Behavior of users in space. You might have noticed how people sit in an unknown place such as a railway station or a park. People avoid sitting face to face with an unknown person, and either sit away or sit with their back to the person (Figure 5.4). On the other hand, if you are going to the restaurant with your friend you would like to talk to her, and hence you prefer eye-to-eye contact and thus sit face to face. Now take the example of group seating for, say, four people. If it is a chatting place like your college coffee shop, then you sit close to one another in a closed group. If your teacher is there, then possibly the three of you would move a little away from your teacher but remain close to one another. So in space there are invisible spaces that people prefer around themselves. When designing such spaces, the orientation of the different elements should be kept in mind. Next time you take up the design of

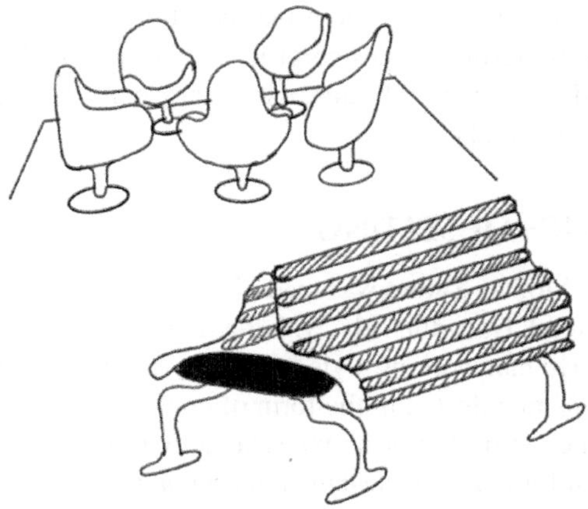

FIGURE 5.4 Behavior of users in space.

a food joint in your town you have to calculate the number of each different seating arrangement based on user behavior at the place.

7. Senior citizens. You should remember that the number of senior citizens (people above the age of 60 years) is gradually increasing. Due to recent advances in medicine, they are leading an active life. While designing the space, these people also need to be considered. For example, with advancing age, the legs start bowing and are no longer straight. Hence, special design intervention is needed to accommodate them in a larger space as they occupy a larger work space envelope. Similarly, these people have flat feet, which means that they require more surface area on staircases. The vertebral column (also known as the spinal cord) becomes "hunched back" and hence they need special backrests in their chairs and seating arrangement.

8. Children and adolescents. Users of this age group have growing bones and have a tendency of getting their body parts trapped in between elements and structures. Thus, at the design stage this age group needs to be considered. The body parts which normally get trapped are the head, neck, and hands. Therefore, the designer needs to be careful not to include any design that could trap the body parts, especially the limbs and the heads of children.

9. Physically challenged. The space should be barrier-free and accessible for the physically challenged so that they can be a part of the common user space and feel comfortable. The design of space should factor in issues like textured flooring, ramps for wheelchair access, and doorways which are broad enough for accommodating this section of the user group.

10. Emergent situations. A normal user under certain situations can no longer function normally. Situations such as fire and smoke, earthquake, and floods are cases where the individual power of locomotion and judgments are lost. You cannot move in a space when the building is shaking in an earthquake. Navigation becomes difficult when there is fire in the space. Imagine how, in spite of the fact that you are sighted, you are unable to move through a room that is full of smoke. Such circumstances need to be factored in during the design process. Thus for such situations in public spaces there should be proper grab rails, emergency exits at frequent intervals, and staircases at regular intervals.

5.2 ERGONOMICS OF SEATING

The body gifted by Mother Nature was made for physical activities like hunting and gathering our own food. Unfortunately, with the Industrial Revolution the demand on humans increased. Human working style and duration both changed drastically. One of these changes was manifested in the form of confining a human to a workstation for a prolonged time, with little or no physical activities. Unfortunately, the human body was not meant for that. It is designed in such a way that physical activities are an essential component for healthy survival. In the world today, humans are spending a major part of their lives confined to a workstation and seated. This is leading to different types of ailments like pain and injury to different body parts. Thus, the need for ergonomically designed seating is in demand. The following paragraphs explain some of the ergonomic principles in seat design.

1. Types of seating. The type of seating is dictated by many factors, and no one seat can suffice for everyone. The type of work the person does plays a very important role. For example, if you are reading then the seat should have a proper back support. If you are a cashier, wherein you are interacting with customers in front and stretching out your hands to give them money or papers, then the seat pan should be inclined a little in front. If you are intending to relax, then a seat with a reclined backrest will be better. If you are doing heavy work like hammering, then a chair with a back-rest is of no use, what you need is just a stool to support your back (buttocks area) (Figure 5.5).

2. Duration. How long you are performing a task is important in selecting the proper seating system. If you are working for 8 hours a day then the seat selection should ensure proper support to the back and the buttocks. If the task that you perform demands that you sit for a very short duration of 10–15 minutes, then a backrest might not be needed or a shorter backrest should be enough.

3. Task. The nature of the task you are performing is important in seat design. If the task demands that you perform some activities in front of you then you have to think differently.

The type of seating in such cases will be guided by the exact nature of the task that you are performing and the design of the work surface in front of you. If you are a cashier then you need to tilt forward and work. In case you are studying then you cannot write with your torso absolutely erect. If you are working at the assembly line then you would probably prefer sitting a little erect.

4. Relevant anthropometric dimensions. Seating design for an individual should take into account a few anthropometric dimensions. Some of these could be seat height (height of the seat), popliteal height (height of the seat pan in front below the knee), buttock to knee length (for depth of seat pan), lumbar height from seat pan (height

FIGURE 5.5 Seating types. (Photo by Trang Doan from Pexels.)

of backrest), elbow height while sitting (elbow rest height), and hip breadth (seat pan breadth). Some more dimensions might be required depending upon the type of seating one has to design (Figure 5.6).

5. *Allowance.* No one sits in one position for long (Figure 5.7). This is true for all types of seating arrangements, ranging from parks to auditoriums. Hence, a change in posture is evident as it also helps in releasing pressure points in the body near the buttock region (and other points of contact in the body) and facilitates blood flow. Thus, there has to be extra allowances in the different dimensions of the seat to account for these different movements. For example, when a person sits then a portion of the flesh of the buttocks and thigh will protrude. The seat pan should accommodate that. Near the buttocks you will find that the flesh tries to protrude and if there are no spaces available then the body near the buttocks would not fit into the chair properly. Over and above you should allow clearance for clothing, especially winter clothing which at times are very thick. If you observe people seating on a chair for a long period of time, then you would find that they changes their posture frequently. Sometimes they sit with knees close and at times with knees spread apart. When you sit with knees apart then you need more space in the seat pan in front. When the person lifts one leg above the other then some extra space is required at the seat pan. Even when you slouch you would notice that your body comes forward and you rest your bodyweight on the buttocks, which slide forward on the seat pan. In case of group seating one needs to calculate the personal space for each individual.

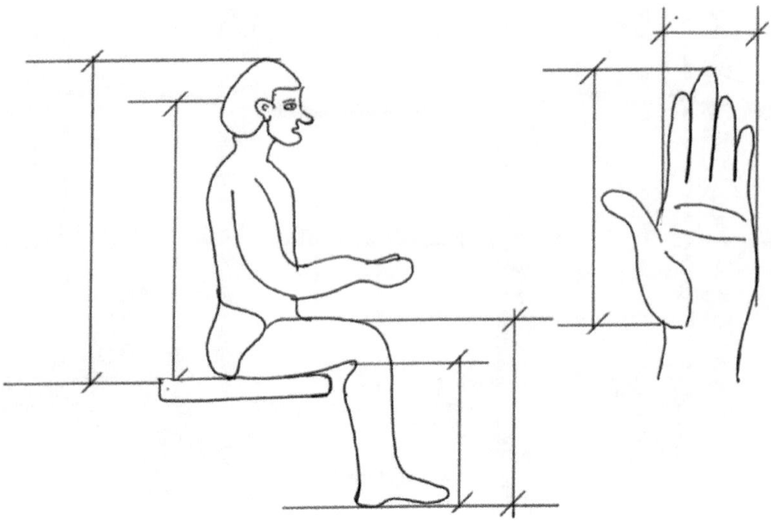

FIGURE 5.6 Anthropometric dimensions required for seating system design.

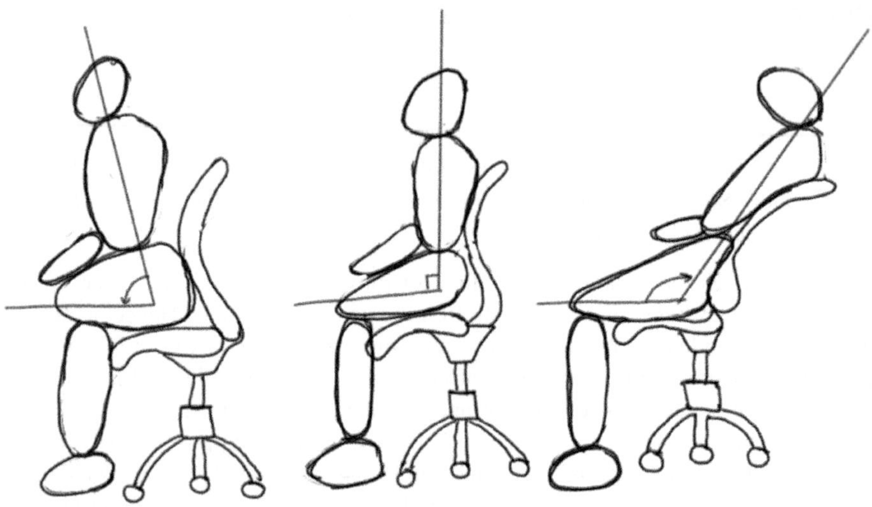

FIGURE 5.7 Allowances required in seating systems.

You might have noticed that when you sit in a park or on a shopping mall bench you prefer keeping some distance from one another, especially if the other person is unknown to you. At times, you may even place a bag at your side to demarcate your "zone" or territory. This is called personal space, which needs to be accounted for in group seating.

6. Backrest. Because the human back is mainly the vertebral column (sometimes called the spinal cord or the spine) and the bones of the shoulder and hip, it at times needs support, especially if one is working for a long time sitting (4–8 hours a day). In such cases, the vertebral column has to remain in what is called a "neutral" position. This neutral position resembles the letter "S" and a somewhat modified form of the letter, which is not that defined but a little flattened at the curves. If not in neutral position then the components of the vertebral column, which are called the vertebrae (bones of the back), are not in alignment and the intervertebral discs which are shock-absorbing cartilages are not compressed uniformly. This leads to disc degeneration and back pain sets in. The situation is somewhat like a vegetable burger. The burgers are the vertebrae, the vegetables inside are the intervertebral discs. If you do not press uniformly on the burger while eating then the vegetables might ooze out from one side. If you apply pressure unequally, the same thing could happen. The same applies to the vertebral column. It has to remain neutral with the bones of your spinal cord compressing uniformly over the cushion (intervertebral disc). To achieve this neutral position the backrest of the seating becomes important. The most important part if you are sitting for shorter work spells, say 15–30 minutes, is the lumbosacral segment. This part is also called the lower back and is roughly the region behind your belly button. This part needs to be supported for neutral posture. Thus, the seat backrest aims to support this part first. If you are sitting for a longer duration, for example, for 8 hours, then the entire back needs to be supported. Support to the head and neck is advised when you need not move your head frequently, as when you are resting on a beach chair (Figure 5.8).

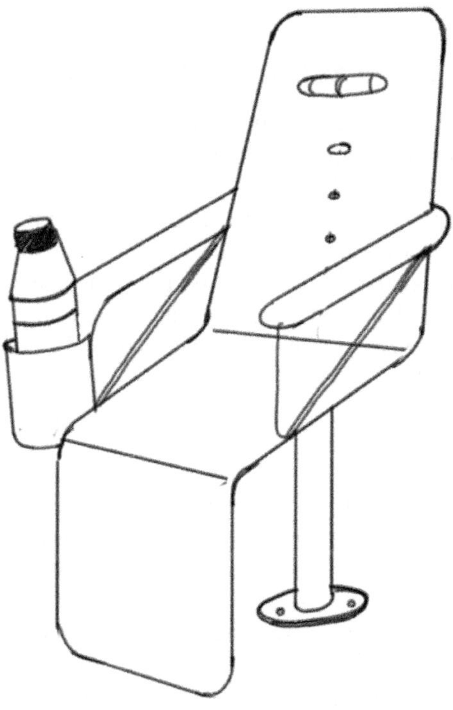

FIGURE 5.8 Backrest, elbow rest, and seat pan padding design.

7. Arm or elbow rest. Whether to provide for elbow rest is decided by whether the task requires moving the torso. If the torso is to remain static then elbow rests are a must.

8. Seat pan padding. Your comfort factor on the chair is decided by the quality of seat padding when you are sitting for a long time on the chair. It is difficult to sit for long on a hard surface as it blocks the flow of blood in certain parts of the thighs and legs, including the buttocks.

5.3 PSYCHOLOGICAL FACTORS IN SEATING

You have heard people talking about lower back pain and the design of a chair with a good backrest. This is what most of us perceive when it comes to chair design. It's important to remember that humans are not designed for sitting but for leading an active life. Hence, if you force humans to sit for long hours then ailments are evident, and no one can help you. Chairs only act as a support to some parts of your body. Lower back pain is one such issue which at times does not have anything to do with the chair design, it is purely psychological. For example, children before examinations have back pain which subsides when the examinations are over. Similarly, people who are stressed due to financial or for other reasons suffer from back pain which subsides after the problems are solved. Therefore, as a designer one needs to know these things so that you do not take the blame for the design, when no design issues are involved in the process.

5.4 BODY STRUCTURES ASSOCIATED WITH SEATING

Your back is the most important structure that is related to seating. Unfortunately, Mother Nature has not provided a guidebook for your back, on how to use it effectively and judiciously so as not to damage it! The bones and vertebrae together make up the vertebral column or the spine. The vertebral column provides support to the back and also aids in movement. It also encases the spinal cord. There are discs which act as shock absorbers in between respective vertebrae. Other than this, there are connective tissues such as tendons, ligaments, and muscles, all of which provide support and facilitate movement of the back. The vertebral column is supposed to be kept at neutral (Figure 5.9).

5.5 DIFFERENT TYPES OF SEATING

There are different types of seating arrangements that we can see all around. Some of these are meant for doing some specific task like reading, cash dispensing, computing, and so on. So, here the seating has to be perceived as a system, and hence has to be considered with the work surface in front. First, we design the work surface and its different dimensions and then we design the chair and fix its different dimensions in tandem with anthropometric dimensions of the body. There is certain seating which is exclusively meant to be used for seating and relaxing purposes only. These are to be designed with a different set of ergonomic principles that should be kept in

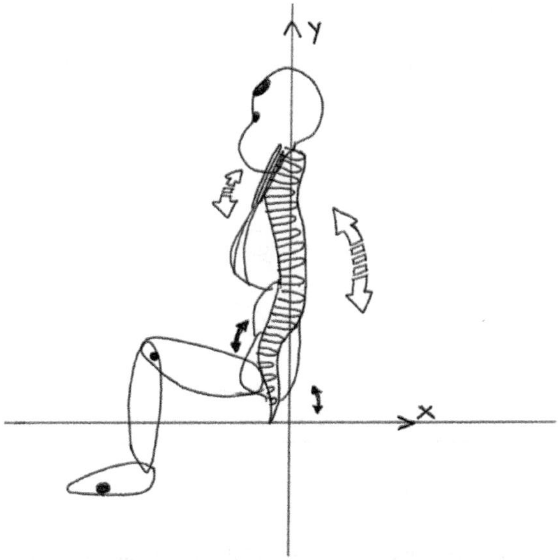

FIGURE 5.9 Body structures associated with seating system design.

mind. Some of the ergonomic principles to be applied in seating design of different types are mentioned below.

5.5.1 Seating Design in General

You have to ask yourself the question: Who is your target user for the product? Are they male, female, or both? Depending on this you are to select the relevant anthropometric dimensions from the database (Chapter 3). Next, one has to equate the different dimensions of the seating with the relevant anthropometric dimensions of the body. The seat height is decided by the sitting height. The seat depth is decided by the buttock to popliteal length. Similarly, for deciding the seat width, two anthropometric dimensions need to be considered: the hip breadth for the seat breadth at the back and the knee-to-knee relaxed position for the seat breadth at the front. This is necessary because when you sit, your knee-to-knee distance increases and this gives you more comfort while sitting. This is because the thigh bones are connected to the knees and the lower limbs. The thighs are connected to the hip bones as well. Thus spreading the knees apart rotates your hip bones to pull the vertebral column in a more neutral position of an inverted S shape and relaxes the back. Hence, you feel more comfortable (Figure 5.10).

Next is the height of the backrest. For deciding this, one needs an insight into the type of task to be performed while seated at the work surface. Along with this, you need to know the duration one has to spend at the specific task. The type of task will dictate the actual design of the chair. For example, if the chair is to be used in front of a computer then it should ensure that the upper and lower parts of the body are maintained at a neutral position. Neutral position means that the vertebral column

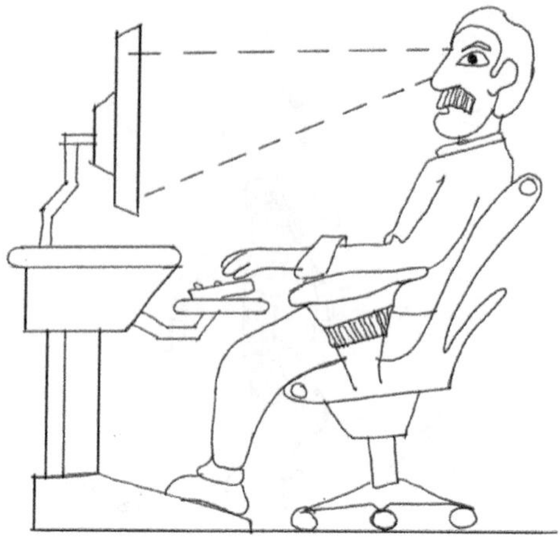

FIGURE 5.10 Seating ergonomics.

should maintain an inverted "S" structure. If the user has to sit and work for longer durations, for example, 8 hours a day, then his entire torso needs support. If the user has to sit for a shorter duration, for example, for 20 minutes at the doctor's clinic, then only supporting the lower back would suffice, including the lumbosacral segment of the vertebral column. The utmost care must be taken when designing the backrest, so that it fits the normal (neutral) contour of the back. As the duration of the sitting task increases the height of the back support should also increase proportionately.

The padding material for seating needs to be considered as through padding one can reduce the contact stress between the body and the chair. You also need to keep the environment in mind. If temperature extremes are anticipated, for example, in an air-conditioned room, then the material for the chair should be a bad conductor of heat, otherwise it will be painful when it comes in contact with the body.

5.5.2 Waiting Areas

Any seating for waiting areas like a doctor's office, office, railway waiting room, will follow the same principles of seating as outlined above. For waiting areas one needs to consider the dynamic movement of each person and based on that the spacing between two chairs should be decided. Often the center table in such areas hits the knee of the users and at times can be very painful. Thus, the height of the center table and its design should consider this aspect of the human body. You also need to keep enough space for every person to move out and into the seat when a group of people are already sitting (Figure 5.11).

5.5.3 Sofa Sets

The sofa is a place on which one normally relaxes at home, in clubs, in a hotel lobby, and so on. It should facilitate the maximum dynamic movements of the body, including sleeping, sitting, sitting cross-legged, and all possible postures. When designing the

FIGURE 5.11 Ergonomics of a seating system in a waiting area.

sofa set, apart from following all the ergonomic principles of seat design, one needs to add dynamic movements like those mentioned above. The padding of the sofa should be of very high quality, neither too hard nor too soft. It should be able to support the different dynamic contours of the body. If the padding is too soft then getting in would be easy, but getting out, wherein one lifts the weight, could be difficult. There should ideally be space for the legs to go back while seated, which facilitates lifting one's weight from the seat while getting up. People often argue about the usage of the armrest in sofa sets. In a single sofa set, armrests provide support to the elbows and reduce stress on the shoulder joints. Thus, seating should facilitate support to the vertebral column and other body members and facilitate dynamic movements (normally made by users when they are on a sofa). The height of the sofa should be in tandem to the popliteal height (height from the undersurface of the thigh just behind the knee) of the user, as well as the seat height (Figure 5.12).

FIGURE 5.12 Ergonomic issues in sofa set design.

5.5.4 Cash Counters

The cash counter is an important place where customers stand on one side and the staff sits on the other; for example, a bank cash transaction counter in a banking scenario. This creates occupational health hazards on the part of the banking staff that are providing the service. The main problem is that the two persons involved are at different levels when it comes to eye and elbow height. Thus, the seated person has to stretch their hands in the upward direction to interact with the customer. The neck is also stretched upward considerably, leading to shoulder and neck pain and other ailments in the long run. This leads to a decline in productivity and the entire services are ultimately affected. To address this issue we need to bring the three anatomical landmarks of the seated and the standing users at par. The solution to this problem could be divided into two parts. One solution could be to raise the platform height on which the staff is seated so that their anatomical landmarks are at a similar level with that of the standing customers. The other solution could be to provide a seating arrangement for the customer, but in that case the counter height has to be reduced to the elbow/sternal notch height of the seated customer. In some cases, the staff could be made to sit on a high seat or stool, otherwise there would be no eye-to-eye contact between the staff and the customers. If both staff and customers are standing then they are at the same level and it is not much of a problem (Figure 5.13).

5.5.5 Transportation Seating: Drivers and Passengers

In the case of a driver's seat, apart from the principles of seating we have already discussed, adjustability is the key requirement. The range of adjustability has to be

FIGURE 5.13 Ergonomics of cash counter design.

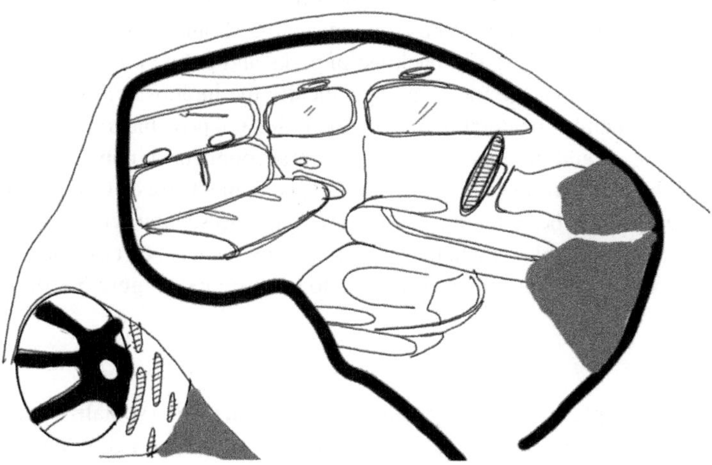

FIGURE 5.14 Ergonomics of transportation seating.

in the vertical direction (seat height) as well as in the horizontal direction for the distance between the different controls and the seat. This is essential, as unlike any other seating, in automobile seating, the driver not only has to be comfortably seated, but the different control elements have to be within easy reach and the display elements should be visible from the specific seating position (Figure 5.14).

5.5.6 Park or Outdoor Seating

The first consideration for any outdoor seating should be the proper selection of material. Such seating systems are exposed to different environmental hazards like

FIGURE 5.15 Ergonomics of outdoor seating.

heat, cold, humidity, rainfall, and so forth, and hence essentially the material should be a bad conductor of heat and at the same time durable. When designing such a seating system you need to focus on the fact that people use them in a number of different ways. In such places, people at times lie down to read books, squat and sit or sit in a cross-legged posture. Thus, public seating should be designed to accommodate such varied postures. The backrest in parks and public places should ideally support the major part of the torso and specifically the lumbosacral segment (lower back) as people spend considerable time on them. It's not possible to provide padding and hence there should be enough allowance to account for dynamic movement of the body so that the buttocks are relieved from the contact stress. Orientation of the seating system should be in tandem with a user study to see the type and number of group formation in the place, whether the groups are in twos, threes, fours or more. Depending on this, the seat dimensions need to be designed (Figure 5.15).

5.5.7 Restaurant Seating

The restaurant is a place where we go not only to eat but also to socialize with friends and acquaintances. Hence, you need to perform a user study to get to know the type and number of group formation as mentioned in the previous section. You need to identify how many users go alone. Based on these observations the seating orientation has to be arrived at. That, is the number of seating for two persons, single person, four persons, or larger groups. A person sitting alone should not feel embarrassed by groups looking at them. Similarly, two unknown persons might not like to sit face-to-face at the table. I was once told that a world-famous fast-food joint employs an ergonomist to design their restaurant furniture in such a way that customers are extremely uncomfortable on them, finish their food quickly and leave, providing the opportunity for other customers to come and occupy those seats. This design intervention ensures a steady flow of customers without any intervention and just by bad ergonomic design! The ergonomic design approach here is designing the seating in such a way that it does not match the anthropometric dimensions of the human body (Figure 5.16).

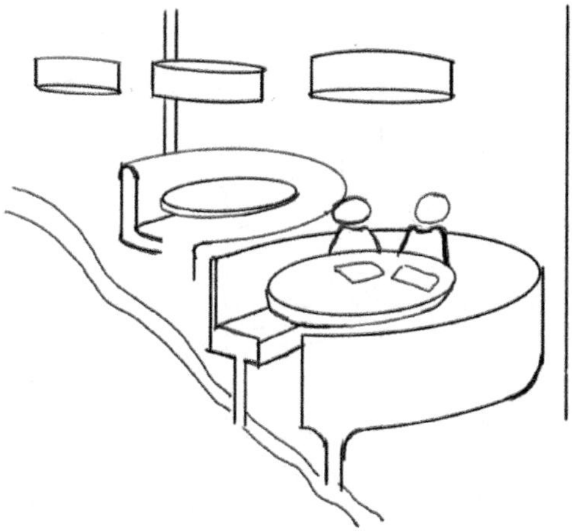

FIGURE 5.16 Ergonomics of restaurant seating.

5.6 SEATING BASICS: NON-DESIGN ISSUES

No one can design a chair which would give the utmost comfort by using it at a stretch for more than 45 minutes to an hour. This is a myth which the designer has to understand. Considering the structure of the human body, no one should or could sit confined to the workstation for more than 40–45 minutes in one position. The body demands a change in posture at frequent intervals, which is essential for productivity and better job output. Thus, one needs to keep in mind the following "softer" issues, often called non-design issues, in designing a seating system.

(a) There is ideally no good or bad posture, what is important is change in posture. The seating system should facilitate this change of posture.
(b) At the computer workstation or for any such activities at a confined position involving the eyes, it is suggested that users take small micro breaks after every 30–45 minutes of continuous work and look at distant objects.

5.7 KITCHEN SPACE

The kitchen is an important work space where the fuel for your body is made. Earlier in many parts of the world, joint families were in existence and that lead to greater demand for food for a larger number of people. Hence, the total area of the kitchen was bigger with country-made ovens mainly fueled by coal or wooden logs. All cooking activities, ranging from cutting vegetables to cooking, were performed while sitting on the floor. This facilitated a change in posture with much more stability to the body (Figure 5.17).

With the passage of time, joint families in these countries have disintegrated and nuclear families have evolved. Unfortunately, amidst this change one thing has remained constant, the food habit. For example, some people still eat a lot (amount and variety of food) for breakfast, lunch, and dinner. Some even prefer eating in

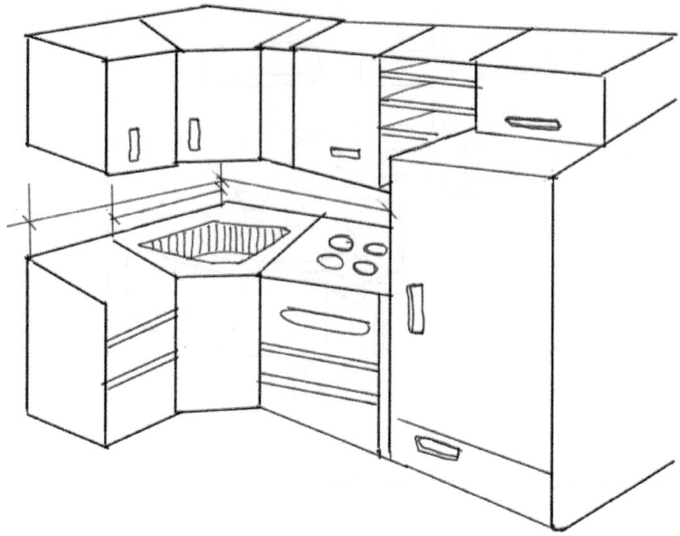

FIGURE 5.17 Ergonomic issues in kitchen space design.

between meals. Cooking so many dishes takes time. If one looks at the compact kitchen today in some of these countries (Asian, mainly), it can be seen to be a western concept implanted into an Asian scenario. The first major change in a compact kitchen is that due to lack of space the user (cook) has to stand and cook; a marked departure from their predecessors. Unfortunately, the demand for food has remained the same or has increased. These people now eat more and hence the extra burden falls on the cook, who has to spend longer in the kitchen while standing and cooking food. This leads to pooling of blood (blood accumulation in both the legs after prolonged standing with minimal movement) in the legs leading to different ailments in the long run such as swelling of legs, varicose veins, to name a few. Thus, we have been gifted with the concept kitchen in the Asian scenario but at the cost of our health.

The main ergonomic considerations in the design of kitchen (compact kitchen) are the dimensions of the space with a focus on elements in the space such as the kitchen platform, cooking space, storage, ventilation, and illumination. In addition, the refrigerator, mixer/grinder, microwave, and similar gadgets also find a place in the kitchen. For deciding upon the height of the kitchen platform one has to keep in mind the height of the gas stove/oven and the utensils to be stored on top of it. The user has to have clear visibility of the bottom of the utensil while standing. The kitchen platform should have invisible areas for preparation (chopping vegetables, applying butter on toast, etc.), cooking, and washing, possibly in a particular sequence. The fridge or the storage should also be an integral part of this triangle. This would facilitate the easy flow of material in the space. Some ergonomic considerations should be taken into account with reference to the space like the height of the microwave oven, which should be such that it does not go above the elbow height of the user, otherwise taking out a hot dish would be a problem. The storage areas above the cooking slab should take into account the vertical maximum up-reach of a lower-percentile person so that reaching the end of the shelves becomes easier. The place for placing the gas bottle should be below the slab but the valve should be within easy reach of the users for easy opening and closing of the gas valve.

In the kitchen, the corners are the most effective and hence need to be effectively used for storage elements. In the kitchen, design-free flow of food handling should be ensured. Proper care should be taken so that stoves and ovens placed above the kitchen platform do not lead to bending or stretching for different users. The flow of materials in the kitchen should not cut the pattern of the work triangle. One also needs to consider sequence of work for right- and the left-handed users. The work area in the kitchen, which starts with the sink on the extreme right-hand side followed by the main work area, cooker, and then accessory work area for putting things down, should never be broken. This would not only increase efficiency but also reduce the onset of fatigue among the users. The triangle in the kitchen mainly comprises the cooker, refrigerator, and the sink, considering the cooker as the cooking area.

5.7.1 Principles of Kitchen Space Layout

The layout of the kitchen is always based on sequence of work without any break to the kitchen triangle mentioned before. For determining the work surface height different ergonomic parameters need to be considered. There are certain tasks in the kitchen which are done a little above the work surface. Such tasks include peeling vegetables, whipping, and beating, and need to be placed a little below the elbow height. Some tasks such as applying butter and chopping are done on the surface and hence need to be placed at elbow height but not below. Some tasks require downward pressure, such as rolling roti or bread, and hence should be placed near the trochanteric height. Ultimately the kitchen platform height has to be decided depending upon the task which is performed for the longest duration. The duration of particular tasks also need to be considered. Tasks done for longer durations will dictate kitchen platform height. The microwave oven should be placed near the eye height and not at the sitting eye height in which the user has to bend a lot.

5.8 BATHROOMS

The bathroom needs to be designed keeping in mind that people expect pleasure or fun here as well as functional efficiency. The placement of the shower, the towel rail, and so on, should all be within easy reach of the users. The toilet which is housed close to the shower is normally the smallest place in the house and is normally neglected. The design of toilets should consider the seating posture. The place for having a shower, which is often demarcated as the shower capsule, is designed by considering the span akimbo of the users. This saves space and thus the design becomes economical (Figure 5.18).

5.9 BEDROOMS

This is where we rest after a hard day's work, when we have nothing to do, or are sick. The bedroom has important elements and one of them is the bed. The design of

FIGURE 5.18 Bathroom ergonomic principles.

the bed should account for the dynamic movement of the human body. One should take the stature, body breadth, span akimbo (distance between the elbows extended to the sides forearm folded and the fingers touching one another, Chapter 3) as reference, but allowances for dynamic movement of the body have to be given. When a person falls asleep there are multiple movements that the body makes. These movements have to be accounted for. The space around the bed has to provide enough room for the person to go to the bathroom at night and drink water. The pattern of walking of humans (called the gait) is a little different at night (somewhat like an alcoholic) and hence enough space has to be given. For this purpose, the hip breadth, maximum body breadth, and the feet-to-feet distance could be taken as reference points, to which allowance could be added. The mattress of the bed should be stable enough to support the body and should not sag, otherwise the person will be uncomfortable. To keep the vertebral column stable, one has to ensure a mattress which is neither hard nor soft, but takes the body contour while sleeping and moving from side to side.

There are certain common movement patterns of the body when a person is sleeping. First is that the person turns the body from one side to the other, and to do that the body rolls in the direction of the turn. Thus, the width of the bed has to account for this movement. Second, the body shifts a little in the upward and downward direction, with the hands moving above the head and this has to be factored in when determining the length of the bed. Therefore, the length and breadth of the bed has to be more than the stature and body breadth to account for such movements. The

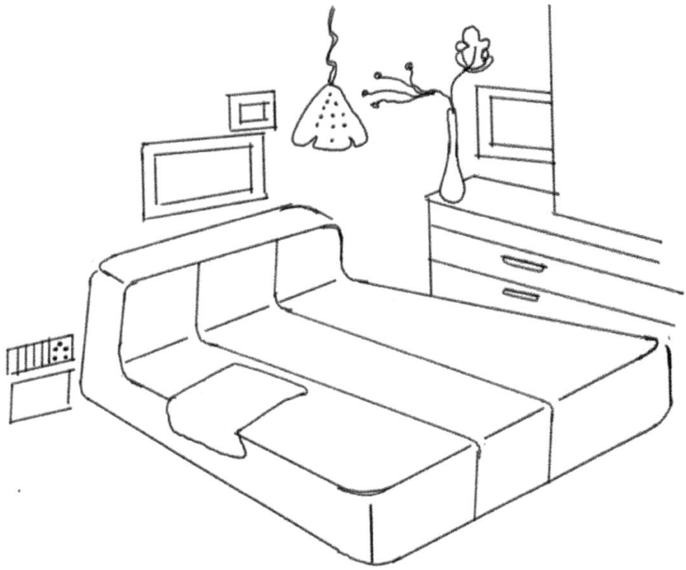

FIGURE 5.19 Ergonomic principles in bedroom design.

height of the bed should be such that a person should be able to access it with ease and normally the midpatellar height (knee joint region) is taken as reference. In any case, the height should not exceed the midpatellar height of the lower-percentile user (Figure 5.19).

5.10 STAIRCASES

The staircase is considered to be the most hazardous of all circulation spaces. People mostly fall at the staircase, and such chances are increased in the case of the elderly and obese users. Serious accidents occur while descending as it becomes difficult to maintain body balance and the person has to halt the body's momentum by grabbing the rails. Thus in staircase design the gradient of the staircase should be factored in so that people climbing them can do so at a relatively low physiological cost without becoming exhausted. Longer step length helps prevent accidents, as the feet have enough room to rest while climbing up as well as while climbing down, provided the staircase is inclined and each stair is not covered by the staircase above. The location of the stairs becomes a problem while coming down the stairs, when the staircases are narrow, and in elderly people who are unable to gauge the next stair compared to the younger population. In staircase design, the direction should be that the height of the grab rail should be at the trochanter height (near the hip), so that the elbow is a little flexed (bent inward) and the person is able to grab the rail firmly. The diameter of the grab rail should take into account the grip inner diameter. Textures on the stairs help in locating the staircase and at the same time increase the friction between the feet and the staircase, thereby reducing the chances of slippage and falling (Figure 5.20).

FIGURE 5.20 Ergonomic principles in staircase design.

5.11 COGNITIVE ERGONOMIC ISSUES IN SPACE

There are many cognitive (Chapter 7) issues in space. Every individual prefers to have a space of his own, in which he does not prefer anyone to come into. This creates an invisible envelope around each person, which is called his "personal territory". When designing space, one has to keep this in mind. When people sit with their loved ones, they prefer to be in each other's personal space. While with an unknown person, some distance is preferred. The dichotomy of human behavior is that they prefer privacy on one hand, and on the other hand they are a social animal and like to be in the company of others. In many offices people demand their own offices completely shut off from others. After some time they realized that it is necessary to interact with people, otherwise the environment becomes very claustrophobic or suffocating. Here came the concept of the "open office" system, wherein small partitions divided a big space into smaller zones. The advantage was that people had their own workstations and privacy, and at the same time when they stood up from their seats they could see their colleagues and feel connected.

Inside small spaces like elevators (lifts) or telephone booths people feel claustrophobic (a sense of suffocation) if it is enclosed on all sides. That's why telephone booths with glass panel which allow a person to stare at the outside world are more preferred, as the act of staring at the outside world acts as a source of resource to carry on the telephonic conversation. This is also called an "attention resource". Similar things happen inside a toilet and hence toilet doors are open at the top with a little gap below the door as well. Inside the elevator, people have similar problems. Such places where it's not possible to provide for an opening could be made a little more relaxing by providing a mirror in which the person is able to see themselves, or by fixing a picture of nature like waterfall, mountain, and so on. These act as sources

FIGURE 5.21 Cognitive ergonomic issues in space design.

of "attention resource" and help people stay comfortable. Light music also helps in such places and it's even better if accompanied by visuals (Figure 5.21).

5.12 RESTAURANTS

There are three major areas in a restaurant (Figure 5.16). The first is the dining area, the second is the kitchen, and the third is the cash counters where transactions happen and orders are taken. For the dining area the furniture elements have to be designed in tandem with the ergonomic principles discussed previously. It has to be borne in mind that at the dining table the plate has to lie in the user's primary zone and this should not be intruded. Second, when a person eats, the two elbows are a little elevated, thus if you visualize this from a plan view and draw a circle, that is the minimum requirement of an individual. Therefore, the table surface should be designed based on this. The chair at the dining table has to be positioned accordingly. Extreme care has to be taken concerning the available space under the table for keeping legs so that they do not touch one another. For this the dimension, "knee extended" can be referred to.

For restaurant seating the job of the designer is to decide upon the number of guests sitting in one direction and the number of guests sitting facing one another, either in groups or just two friends. This could only be arrived at after a careful user study to see the number of people coming to the restaurant alone and in different groups. Those coming alone would prefer not to face other people. Those coming in groups would prefer sitting face to face for proper interaction.

The space inside the dining area will be governed by the movement patterns of the staff and customers as well as crowd formation and dispersion during peak times. This has to be studied in great detail to come to the exact layout of the space.

The design of the kitchen will follow similar principles as discussed before. Only the flow of people would increase considering the volume of the customers and the number of waiters. There is a similar case for the cash counter, which follows the same ergonomic principles as applied in the case of counter design.

5.13 AUDITORIUMS

The auditorium or cinema hall is a place where there are a couple of ergonomic issues involved. To start with, the ticket counter should be designed as per the ergonomic principles of counter design discussed before. At the counter, apart from the height of the counter which has been discussed before, a recess could be provided so that the person can move closer to the window, and even if people from the sides try to move closer to the window, they would remain away due to the "bulge" at the counter.

For deciding upon the opening of the counter through which transactions of money and tickets would take place, the first circumference could be considered to be the reference anthropometric dimension, to which allowances should be added depending upon the range of dynamicity. Care should also be taken to ensure that the person does not put his hands through and access the table inside the counter. Communication at the counter is an important attribute and hence the opening on the glass partition should enable a person to talk with ease without bending their body. If they have to bend their body to bring their lips near the opening then the queue behind them would be pushed backward, creating chaos. Thus, the opening should be either a longitudinal slit which can accommodate the lip height of different users, or should be designed keeping in mind taller and shorter users, in which case a circular opening for communication should be provided.

The most important issues in an auditorium are the entrances and exits. Based on crowd flow inside the space, it's imperative that the entrances and exits are positioned accordingly. To determine the door width and height two landmarks are essential. For height, stature (normal) could be referred to. For the width of the door, body breadth or the elbow-to-elbow relaxed distance of users could be taken as two of the many reference points along with others allowing for dynamicity in movement. One has to remember that the crowd is organized when the show starts but gets unorganized when the show ends as everyone is in a hurry to go home.

Seat design for the auditorium has to take into account the utmost comfort of people and should be in tandem with the ergonomic principles discussed before for seat design. The most important issue inside an auditorium is visibility of the screen or stage for all audience members from each part of the space. There is every possibility that a taller person might occupy the seat in front and the shorter person a seat just behind them. In this case, the taller person gets in the way of the visual angle of the shorter person. When designing the seating system and the angle of the gallery one has to be careful to ensure that the visual angle both above and below the horizontal is not obstructed for a shorter person sitting behind a taller person. First you decide upon the inclination of the gallery, taking into account the visual angle (vertical and horizontal) of each user. For this exercise, the slope of the gallery has to be decided, keeping in mind the seated eye height of the lowest percentile and the highest percentile users. If the visual angle of each user is drawn (both horizontal and vertical) with reference to the horizontal line (approximately from the eye height) it will be clear in what manner the visual angle of the shorter person is being impaired, and the slope of the gallery or the seat heights can be adjusted accordingly. In general you can draw the vertical visual angle at 50–60° above the horizontal and 70–80° below the horizontal without moving the head. Similarly, the horizontal visual angle for both eyes without moving the head is roughly 180° and this reduces to 140° in the elderly. The horizontal cone of vision also needs consideration to ensure that the screen and the stage are visible to all. If you draw the vertical and horizontal visual angles from the eye height (while seating) for different percentile values it will provide a very rough estimation of the placement of the screen and the stage even if a taller person sits in front of a shorter person. This calculation of the visual angle (just an approximation!) is not that simple and involves complex calculations as it is also dependent on the distance of the user from the stage or the screen and the size of the stage or the screen. If a person is seated very close to a large stage or screen then it would be very difficult to view it (Figure 5.22). You also have to account for movement of people between seats in the rows because in auditoriums seats are not filled at the same time. It is possible that there are people sitting on the sides and few at the center and people need to move for various reasons. So there should be enough space between the backrest of the front seat and the knee of the seated person so as to allow for the movement of people. For this purpose, one could consider the circumference of the knee of the standing person and the feet length so that the person could just move their body keeping their feet firmly on the ground and without hitting

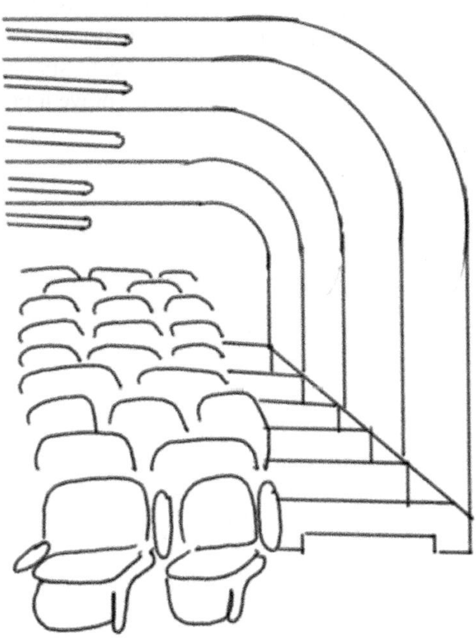

FIGURE 5.22 Ergonomics principles in designing an auditorium.

the legs of those already seated (anticipating that seated people would draw their feet under the seat when these people cross them. This is only to allow the person to enter the row to take their seats facing the screen (bare minimum space). If more space is to be allocated then other dimensions like elbow-to-elbow relaxed or span akimbo (Chapter 3) could be taken, although that would be too much of a luxury and involve wastage of space.

5.14 CLASSROOMS

The main purpose of a classroom is to facilitate learning by making it a fun element. The size of the classroom is determined by the number of students and the type of furniture being used, which together makes up the personal territory of the individual user. Similar to auditorium design, the blackboard needs to be positioned with the visual angle of users in mind. The furniture design should follow similar principles of seating ergonomics, and should ensure proper support to the back and facilitate change of posture, with adequate clearance between the body and the furniture elements. The teacher and whatever they write and project should be visible to all students and this is where the concept of a gallery in a classroom started, where a large number of students have to be addressed. There should be enough space between the benches for the students to move and occupy their seats, for which the body breadth and hip breadth could be considered (Figure 5.23).

5.15 BUS STOPS

This is a unique space where people stand/sit for a transient period. The typical bus stop comprises a good seating system, not so luxurious that it is misused for gossiping. So, the ergonomic strategy here would be to support the buttocks and relieve the muscles of the lower part of the body, and prevent pooling of blood. The

FIGURE 5.23 Ergonomic issues in classroom design.

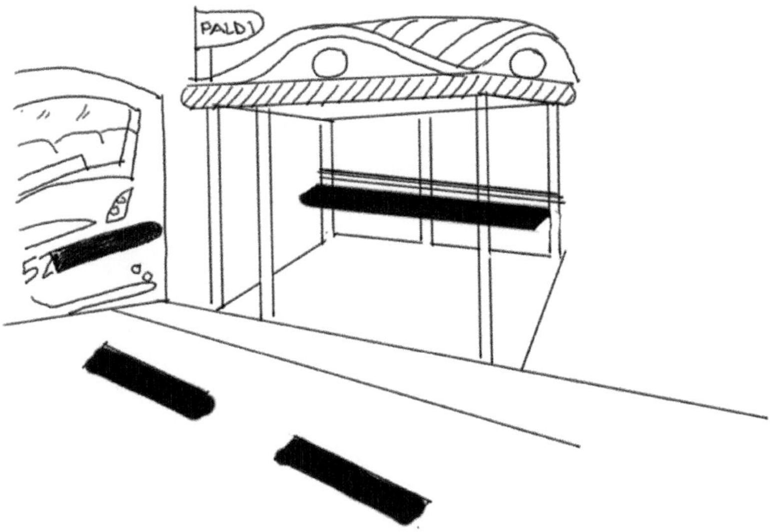

FIGURE 5.24 Ergonomic issues in designing a bus stop.

bus stop is also a shelter for people against environmental hazards such as heat, rain, and cold. In a tropical country, the bus stand should always be kept open facilitating proper ventilation. In cold areas, enclosed bus stops are preferred to insulate and maintain a comfortable temperature inside the bus stop (Figure 5.24).

5.16 ASSIGNMENTS

1. Select a roadside telephone booth and do an ergonomic analysis of it. Try to perceive the space in a different perspective and visualize it from the viewpoint of the users.
2. Conduct a detailed ergonomic analysis of a ticket counter at a train station. Focus on the cognitive ergonomic issues at the counter as well.
3. Select any restaurant of your choice. Observe the movements of people in the space, including the customers and the staff. Identify ergonomic issues in the space.
4. Conduct an ergonomic survey of the kitchen at home. Identify different ergonomic issues such as height of the kitchen platform, space for movement, storage, and washing zones.

6 Exhibition Ergonomics

OVERVIEW

This chapter explores different ergonomic principles in exhibition design, ranging from a museum to an outdoor fair. The issues of crowd flow and control, placement of exhibits, and the pattern movements of crowds are discussed.

6.1 A CROWD AND ITS MANAGEMENT

Exhibitions or displays are crucial as they aim to showcase products to people. All the ergonomics principles should be applied to displays, especially those related to space, product, and cognitive ergonomics. This area of ergonomics focuses to a great extent on crowd flow and crowd management because we have to deal with large crowds as designers in this domain. As exhibition designers you have to deal with large crowds and manage your display with expertise so that everyone is able to appreciate it. A crowd is defined as a congregation of people who assemble at a place for a common purpose. It's a generic term, but we will use it specifically in relation to exhibition design. A crowd typically comprises people of different dimensions, cultures, strengths, cognitive attributes, and so on. Therefore, a crowd in the context of design is a heterogeneous group of people each with a common purpose but with different abilities. It could be a crowd inside a museum, a crowd in a football stadium, or a similar crowd at an outdoor fair.

Crowd management can be done in many different ways. We should first try and educate the crowd that they should be organized, refrain from breaking the rules, and so on. If this does not work then ergonomic interventions like "design" to control the crowd should be explored. The placement of a railing or a barricade would automatically make people form a queue and hence an unorganized crowd could easily be organized. As ergonomists and designers our focus should be on aspects of crowd control through design intervention. At this juncture, the anthropometric dimensions of the crowd need to be factored in to calculate the height of the barricade, so that people are not able to cross it. You also need to consider the strength (push, pull, grip, torque, etc.) of the crowd to decide on the materials and form of the barricades used in situations for crowd control so that they do not break down (Figure 6.1).

6.2 CROWD MOVEMENT

As we have learned in anthropometry, each person has their own personal space or envelope. When the person moves, this envelope also moves. Thus, crowd movement could be visualized as the movement of different sizes of envelopes in space depending on the anthropometric dimensions of the users in space. A critical problem that arises at this juncture is that each individual walks at their own pace, so if the person in front

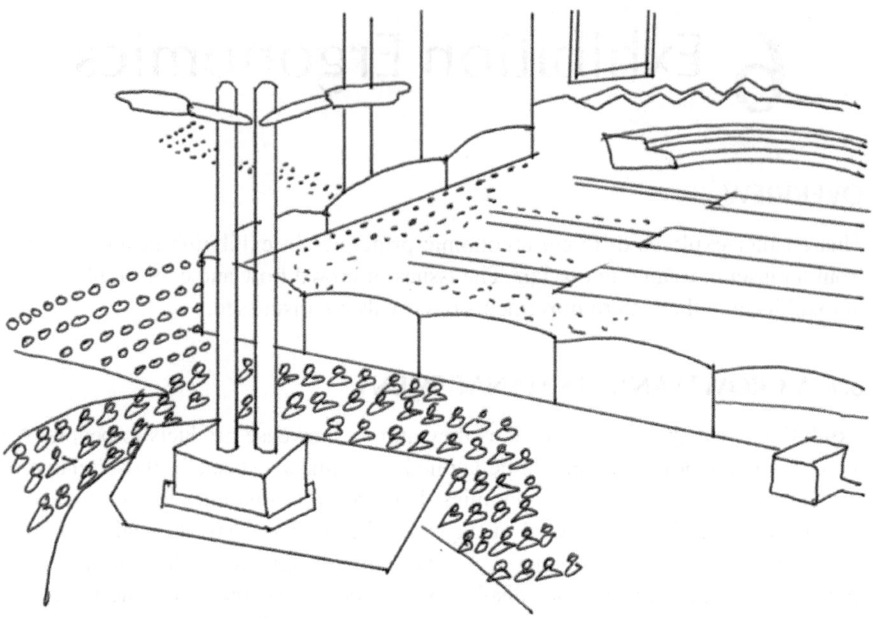

FIGURE 6.1 Crowd flow and its management through ergonomics.

is walking too slowly and there is a huge crowd behind, then the entire crowd gets staggered. Similarly, when a large crowd moves there is always a ripple effect of a push being generated on the people in front if the people in front are not moving as fast as the people at the back. This often leads to stampedes at large gatherings. The problem here is that each person moves at their own pace, some very fast and some very slow. If people at the back try to move faster than those in front, then the crowd becomes staggered and the overall movement of the crowd is slowed down. Thus, the faster you try to move in a crowd the slower your pace becomes. When designing the space for exhibits we have to remember that that each individual has their own characteristics. Some take more time in looking at exhibits and some less time. Therefore, the movement of individual envelopes is different. Thus, the crowd can stagnate if there's not enough space available for people to overtake and move along.

6.3 COGNITIVE ERGONOMICS IN EXHIBITIONS

In any exhibit people need to have maximum information about what is going on inside before they move in. Users also need to know what is located where. Normally by stereotype, people have a tendency of taking a clockwise route, starting from their left at the entrance of an exhibition and exiting from the right. Therefore, information has to be strategically located keeping this attribute in mind. The most important things in space such as facilities and amenities need to be clearly indicated with exits at frequent intervals. This is a requirement for users because they frequently hesitate to enter a big exhibit in case they have to see the entire arrangement, so providing

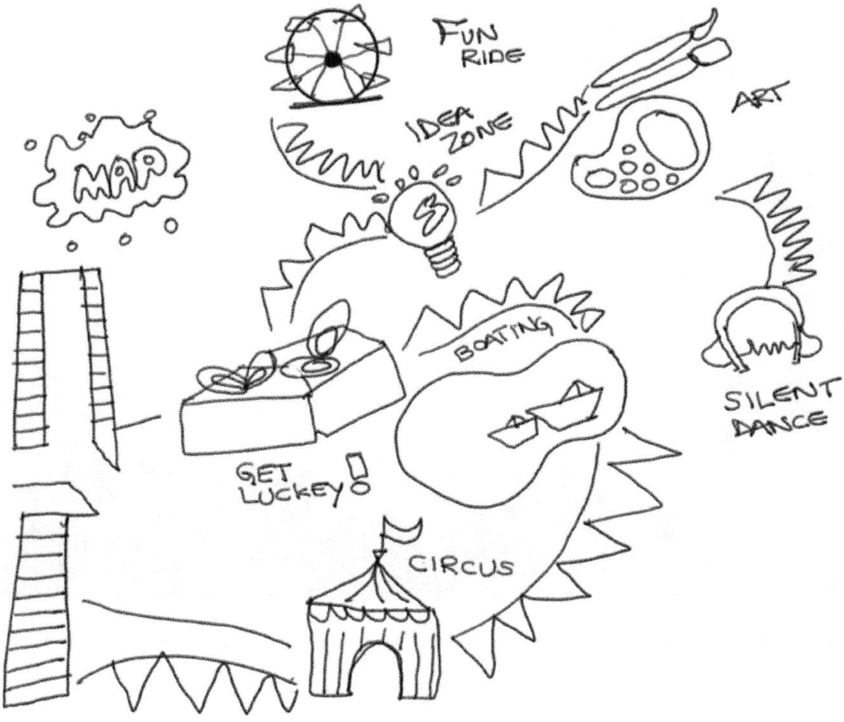

FIGURE 6.2 Cognitive ergonomic issues for users while moving in space.

exits for users is a way to invite them to the exhibition. Seating arrangements at regular intervals are also a requirement to allow fatigued users to rest before they continue. In the absence of these, users might leave the exhibition or might not enter the exhibition hall at all (Figure 6.2).

6.4 PLACEMENT OF EXHIBITS

The exhibits that will be placed need to focus on two attributes. The first is the exact location and the second is the distance from which you want your user to see the exhibits. It is important that any exhibit should be placed within the visual cone of the user from the position where they stand. The normal vertical visual cone of a person without moving the head is at an angle of approximately 50–60° above and 70–80° below the horizontal. Similarly, the normal horizontal cone of vision of use without moving the head is around 110°. Hence, it's important to keep this in mind when placing the exhibits (Figure 6.3).

6.5 DARK AND LIGHT ADAPTATION IN EXHIBITS

The amount of light entering the eye is controlled by the pupil or aperture in the human eye. In intense light the pupil becomes small to permit less light and in low

FIGURE 6.3 Ergonomic issues in the placement of exhibits. (Photo by Hiếu Vương from Pexels.)

illumination or dark environments the pupil dilates to facilitate more light so that one is able to see clearly. When a person enters from a brightly lit area to a relatively dimly lit area it takes time for the pupil to dilate (called dark adaptation). When the same person moves from a dark area to a brightly lit area the pupil takes much less time to constrict (called light adaptation). These adaptations are also facilitated by an adaptation by the nerve cells in the eyes and the brain. Along with this there are certain chemical conversions of photo chemicals of the eyes which also aid in this process of light and dark adaptation. This phenomenon has a design connect in that when you are designing exhibits in enclosed spaces, which need to be artificially illuminated, then one has to keep in mind that a person who enters the space suddenly from bright sunlight will not be able to see things properly for at least 20–25 minutes, so you need to give time for the eyes to adapt. At many venues, graded foot lighting is provided, the intensity of which gradually increases as the person walks into the space. This is to compensate for the dark adaptation of the eyes (Figure 6.4).

6.6 MOVEMENT PATTERNS IN SPACE

Some people prefer reference materials on the left, or prefer keeping to their left as a result of population stereotype (Chapter 7). Thus, when designing spaces for exhibits one should design in such a way to respect this phenomenon. For example, irrespective of the exhibit space shape, any user would generally start viewing the exhibit in an anticlockwise direction (from their left) and then come out from the

FIGURE 6.4 Dark and light adaptation in space design. (Photo by Anthony Macajone from Pexels.)

space. Therefore, exhibits should be arranged sequentially if they are linked to one another to help build the story line of the exhibition.

6.7 ASSIGNMENTS

1. Identify the ergonomic issues in placing a small stone statue in your room so that visitors are better able to appreciate it.
2. Do an ergonomic analysis of a village fair and suggest an ergonomic solution for designing the different stalls and the placement of different merchandize so as to grab the attention of the prospective buyers.
3. You are asked to do an ergonomic consultancy for a supermarket. Discuss the steps you would follow to do this.
4. Do an ergonomic analysis of an auditorium so as to ensure every audience member is able to see the performance on the stage irrespective of their stature.

7 Cognitive Ergonomics

OVERVIEW

This chapter introduces the reader to cognitive ergonomics. It highlights the different elements of cognitive ergonomics such as human characteristics, information processing, mental models, stereotype, and compatibility, and their application in design.

7.1 COGNITIVE ERGONOMICS IN DESIGN

To the designer, the term "cognitive ergonomics" looks like technical jargon. In simple language, it means the information processing (Figure 7.1) and decision-making capabilities of the human brain. You can also visualize it: when information from the world around enters the brain through the eyes, ears, and so on (these are the channels through which information enters the body) it is analyzed, and the brain instructs different parts of the body to act accordingly. These entire series of acts by the brain is called cognitive ergonomics. Technically, it's the information processing and decision-making attributes of the human brain along with some other characteristics typical of humans. To simplify this subject, this chapter has been divided into principles, human characteristics, information processing, mental model, compatibility, and stereotype.

7.2 PRINCIPLES

There are certain principles of cognitive ergonomics which can be applied at the design stage itself to increase the product/space acceptability by your users (Figure 7.2).

i. *Action in relation to your thoughts.* The action at any product interface that you perform should be as per your thoughts or expectations. If you press the switch for the light, the fan should not turn on; this action would not be in accordance with this principle.
ii. *Feedback.* All actions that we perform should give us feedback and tell us exactly what is happening. For example, you go to the ATM to withdraw money. You insert your card, key in the secret number (known as the PIN), and then nothing happens on the monitor for, say, 5 minutes. You are worried and start banging on the machine and finally ask the people around to help you out. Someone does this task and tells you that this machine takes some time to dispense cash. If there had been an indication on the monitor "your request is being processed, please wait" you would not have panicked.
iii. *Use of your sensory channels judiciously.* To provide feedback you use any of the channels in the body through which information can enter. These channels are the eyes, ears, nose, tongue, touch, etc. and are often referred to as the "modality" or "window" through which information enters the body.

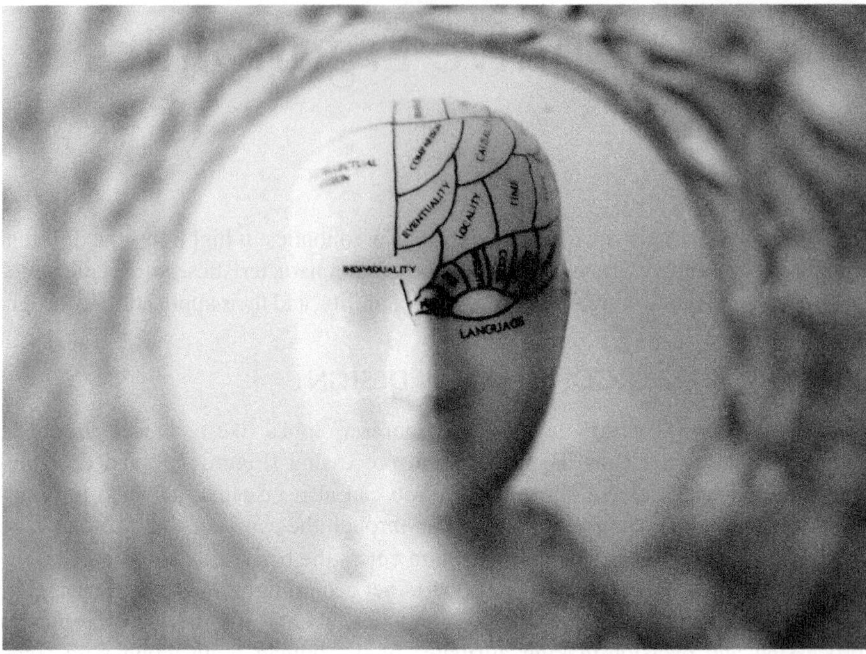

FIGURE 7.1 Cognitive ergonomics. (Photo by Meo from Pexels.)

FIGURE 7.2 Different cognitive ergonomic principles. (Photo by Francesco Paggiaro from Pexels.)

Have you seen an ambulance on the road? It uses a beacon of light, which flickers, and there is also a siren, the reason being that if someone fails to see the light (in case they are looking in the other direction) they would be able to hear the sound. After hearing the sound, they would be alerted. So at times, to provide effective feedback you might have to use more than one sensory channel.

iv. *Use "useless information" at times.* If you provide both the postal code and the name of the post office for a letter at the same time, then apparently the post office name is useless information. This is because the postal code gives all the clues to the jurisdiction under which post office it falls. Imagine a situation where, because of rain, the postal code is erased and is not legible. Now your letter would not reach its destination. If you had written the name of the post office then there was at least a chance of it being delivered at the correct address. In critical operations such information becomes useful. While using certain drugs, the nursing staff in the Intensive Care Unit (ICU) at a hospital has to be extremely careful while injecting such medicines. A minute over-dosage could lead to death. In such cases, devices are designed to give feedback in the form of visual cues, auditory (sound) cues, and also two to three levels of warnings like "you are injecting _ml drug, proceed?" This is because under mental stress the staff should not commit a mistake which could cost a precious life.

v. *Vary the presentation of information.* When the same information (either sound or light) is used for providing feedback to the user, sometimes it is not effective in humans because they get used to it and start ignoring it. This is why the ambulance siren (sound) is not constant but variable (increases and decreases) so that it is identified among the other noises on busy roads.

7.3 HUMAN CHARACTERISTICS

As humans, we have certain inherent characteristics, which are more or less constant globally. One needs to know these so that at the design stage we are able to address them and make the design much more humane (Figure 7.3).

i. *Mistakes.* You have heard the old saying "to err is human and to forgive is divine". Humans make mistakes. This is their inherent property. You have to keep this in mind when designing. We have seen at the systems level how a human reduces the overall reliability of a human–machine system. By design, we need to devise some features so that the chances of making mistakes are reduced. This could be achieved by incorporating feedback and reconfirmation in the system, or designing in such a way that the users are forced to work in certain ways. Even after that, if mistakes happen, guide your users how to come out of that mistake.

ii. *No long routes please!* We have a preference for shortcuts. Be it an examination or traveling through a field, we always look for it. This at times becomes handy for expert users of a device. So, we have designed "shortcut" keys for our computer keyboard.

FIGURE 7.3 Different human characteristics.

iii. *Expect feedback.* We always expect feedback. When you switch on the fan, the first feedback is the click sound of the switch. So even in dark you get to know that you have switched it on in the correct manner. You are using an electric iron for ironing your clothes. You plug in the device, the LED glows, and you adjust the setting to the cloth type, for example "silk". Because the device is regulated by a thermostat, you will only start using the iron when the red LED does not glow any more "probably" indicating it's hot (how hot you do not know, until you touch it). The problem starts from here. As you iron you touch the baseplate to check how hot it is as there is no direct feedback about the "rate of cooling" of the iron. An LED strip on the handle could probably solve this problem. Let us say that we design the LED strip in such a manner that when the iron is completely heated the entire strip is illuminated, and the length of the strip decreases as the iron gets cold. This gives the users direct feedback about the rate of cooling of the iron, so that one could regulate the speed of ironing looking at how fast or slow it is cooling down.

iv. *Reconfirmation.* You are going home for your summer vacation. You go to the railway station and hear an announcement that the train for your hometown is cancelled. Your immediate reaction would be to ask someone whether what you have heard is true. In the process, you end up asking a number of people on the platform. Only when you are satisfied do you leave the station. In essence, you were "reconfirming" the news of the train cancellation. This is another human characteristic. When you try deleting a folder, have you noticed what it prompts? It asks "Are you sure you want to delete?" If you say yes, there is another level of warning "this will delete the folder and all its contents". The system is reconfirming with you, because the consequences of this act could be disastrous. So, a human requires reconfirmation as a feedback to a particular function.

v. *Developing a mental model.* Humans develop a mental model when performing any task. You can navigate to the bathroom of your house at night without switching on the lights. This is not possible when you go to a new place.

Similarly, you can drive your car at ease without having to look at the controls. When you started learning to drive, you had to look at the controls while driving, especially the gear shift.

7.4 SENSORY CHANNELS

You have antennae for capturing signals transmitted by your satellite television service provider. Similarly, the body is provided with such antennae called "receptors", meaning that which receives information from the outside world. Some of these receptors are also called sensory channels. They are the avenues or windows through which information enters the body (brain) for processing and subsequent action. There are generally five types of receptors or channels in the human body. They are located in the topmost part of the body and close to the brain (except touch), so that communication becomes fast and efficient. These receptors, also called special senses, are the eyes, ears, nose, tongue, and touch.

7.4.1 EYE

The structure of the human eye is very important in design and can be compared with a camera. In layman's terms, it comprises different parts. The frontal part of the eye has an opening called the pupil. This part contracts and dilates to adjust the amount of light entering the eye and falling on the photographic plate called the retina. Unlike the camera, through chemical reactions in the retina images are formed and are sent to the brain for interpretation. Behind the pupil lies the eye lens, which is a jelly-like structure. It helps in focusing the image of distant and near objects (Figure 7.4).

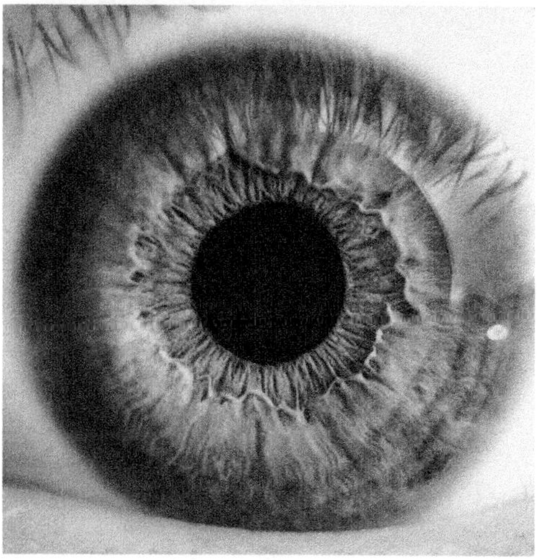

FIGURE 7.4 Structure of the human eye and its relevance to design. (Photo by Tookapic from Pexels.)

Design pertinence. The retina is made up of cells called cone cells which help in color perception. Globally, approximately 6% of the male population is red and green color blind, which means they cannot perceive pure red and pure green but instead perceive them in shades of gray. Less than 1% of the female population suffers from such color blindness, so women should be happy! Similarly, elderly people above the age of 60 cannot perceive pure blue and hence perceive it in shades of gray. This is due to the degeneration of the receptive cones for that particular color. Thus, the solution to these problems could be to change the color spectrum from pure to a little impure by mixing in some additives. For example, pure green can be made yellowish green, which is then visible to all. Similarly, pure red can be made amber red. Pure blue can also be made whitish blue. These colors then become visible to all.

7.4.2 Ear

The ear (Figure 7.5) is a device to receive sounds from the external world. The ear in humans, unlike those of the four-legged animals, is not able to make movements on its own. However, it is a direct connection from the external world to the brain.

Design pertinence. The ear acts as a reconfirmatory channel for information reception along with the eye. That is why we use audiovisual techniques for communicating information more effectively to users. At times in human environments, for example

FIGURE 7.5 Structure of the human ear and its relevance to design. (Photo by Aleksandr Slobodianyk from Pexels.)

brightly lit areas, it is not possible to communicate anything through visuals. Here, the auditory mode of communication works well. In cases where visual channels are overloaded with information, such as in a nuclear power plant control and display, auditory channels are used to convey warning messages. Audiovisual outputs are at times reconfirmatory in nature and substantiate what the eyes have seen or the ears have heard. In a noisy environment auditory signals can be made audible if their frequency of transmission is changed. This is exactly what happens when an ambulance travels on a noisy street with the siren on. The siren does not emit a constant sound, but a sound with changing frequencies.

7.4.3 Nose

The nose is an organ which receives smell (Figure 7.6) of different types and kinds from the external world. Although the capacity of this sense organ in humans is limited compared to those of animals, it nevertheless plays an important role in appreciating the food that we eat and also different types of aromas around us.

Design pertinence. Household LPG gas does not have any odor. It is mixed with an odor so that users are able to detect its leakage. Similarly in tunnels where sound or light cannot be used, olfactory channels are used by releasing a particular odor in the tunnels to warn the occupants of a disaster and to vacate the tunnel immediately.

FIGURE 7.6 The human nose and its pertinence to design. (Photo by Alexander Krivitskiy from Pexels.)

There were clocks which were used in olden times which burned an incense stick of a particular perfume when the hour hand of the clock reached certain clock positions. Thus, the users would get an idea (reconfirmation) through sound and smell the exact time of the day. Even if they missed the time the fragrance in the air would help them recollect the time span which has just elapsed. Odor is used to advertise certain toiletries like soaps. Through a special technology, the soap's perfume is impregnated on photographs of the soap. When the user rubs their fingers on such photographs, they would smell the exact perfume. Such olfactory senses are used to create ambience within spaces. For example, special room sprays are used within some aircrafts or restaurants which are very specific for those places and not normally available on the open market. This particular essence generates a mental model in the users and hence in this way the company tries to build up their brand identity among the users. The next time they enter such a space the recall value of such an essence brings back their previous experience in the same space.

7.4.4 Touch

Just like the antenna on numerous devices, there are numerous antennae all over the skin, and some are very sensitive in certain locations of the body on the surface of the skin. These antennae are known as senses or sensory receptors, as they receive information from the outside world and transmit them to the brain. The senses which are sensitive to touch are also called the tactile senses. In general we use touch or tactile senses very grossly. The sensation of touch can be categorized into haptic senses (sensitive to pressure) and kinesthetic senses (sensitive to movement in space). There are some receptors which are responsive to heat (Ruffinis end organs), cold (Krause end bulb), and pressure (Pacinian corpuscles).

Design pertinence. When designing control knobs either for a motorbike or a cellular phone, the tactile receptors come into play. The better the recognition by the tactile receptors of the palms, the faster and more comfortable is the response. When it comes to textured controls, the tips of the fingers are very sensitive to them. You may try this experiment at home. Try running your palm over a textured surface and you will notice that compared to the other areas of the palm, the fingertips are very sensitive in detecting them. This is because of the presence of a special category of receptors at the tip called the Iggo dome receptors, which are extremely sensitive to textures. Thus, depending on the overall position of the hand and the palm, and the reachability of the fingers, the textures on the control switches need to be developed so that the Ibido receptors are in contact with the textured surface (Figure 7.7).

7.5 INFORMATION PROCESSING

We have seen that information from the outside world (Figure 7.8) has to enter the brain, which is where it is processed. This information enters through different channels, or modalities as they are also called. The senses and the sense organs are some of these channels. The entire information processing takes place very quickly and is very complicated, but to understand it we break it up into the following five stages.

FIGURE 7.7 Human touch and its pertinence to design. (Photo by Fancycrave.com from Pexels.)

FIGURE 7.8 Human information processing. (Photo by Rafael Cosquiere from Pexels.)

I. Sensation. This is the first stage of information processing, whereby information from the outside world is channeled through the respective sense organs. For example, if there is a fire in your room, what happens? The first thing that you notice is the light and sound of the fire alarm and if you are close or in front of the alarm. So the sound of the alarm or the light has to enter your brain through the ears or the eyes. This is the first stage of information processing.

Design pertinence. If this stage of information processing is taken care of through design then the other stages follow automatically with relative ease and less error. For example, a friend of mine left a voice message on my landline phone. The next day although I entered my office, I did not make or received any calls. Later, I came to know that I lost a very big order, the information for which my friend had left over the voice message. Incorporating a small LED, to flicker and indicate that a voice message is there, could have solved the problem from the viewpoint of drawing my attention through my "visual" senses.

II. Perception. This is the second stage of information processing. The user in this stage tries to perceive or understand what the information is all about. On hearing the fire alarm the user understands that the alarm that they have heard or the light that they have seen is indicative of fire in the room.

Design pertinence. Sensation and perception go together and the design interventions for both stages are similar. The example of the LED in the previous stage also takes care of sensation and perception attributes. In fact, when sensation is taken care of then good perception sets in.

III. Cognition. This is the third stage of information processing and is often called the "heart of information processing". This is when the brain synthesizes and analyzes the information which has already entered. It is this stage where there cannot be any direct design intervention. If the previous two stages are taken care of through design, then cognition sets in.

IV. Decision making. This is the fourth stage of information processing, during which the brain makes a decision. In case of fire, the brain taking into account the gravity of the situation makes a decision that one should immediately run out of the room.

Design pertinence. In a telephone, one can facilitate the decision-making attribute of the users by providing all the possible options. Depending upon the options the users need most (through user studies) such as redial, speed dialing, and recall, the different options could be included and arranged to facilitate users' decision making.

V. Action function. As the name suggests, this is the last stage, where the user physically acts or makes a movement. This is the stage where physical ergonomics such as anthropometric dimensions, strength, and so on, need to be considered. So at this stage (in case of fire in the room), the user actually runs out of the room.

Design pertinence. The size, shape, and clearance of the buttons of the telephone are important elements to be considered at this stage. Anthropometric dimensions of the hand along with strength data are important elements to be considered at this stage.

7.6 MENTAL MODEL

Imagine the last time you went to a cinema to watch a movie. You can recollect along with the movie, the cinema and its structure and design. Next time when someone tells you that he has been to a cinema, even if you have not visited that particular cinema, an image of a cinema appears in your mind and you can understand what a cinema is. Have you ever considered that when someone talks about a cinema you do not visualize an animal like an elephant? This is due to the fact that anything we are exposed to in our surroundings leaves a mark in our brain and this is called the mental model. When someone tells you that apples are sweet you do not visualize a piece of chicken but an apple, because you have seen an apple before (Figure 7.9).

There are two types of mental models. One is static and the other is dynamic. As the name suggests, the static mental model is that model which is stationary or at rest. For example, the icon of wastepaper basket on your desktop is associated with waste or junk papers. So the trash can (trash or recycle bin) icon is an example of a static mental model. On the other hand, a dynamic mental model involves movement and is the model that we develop while performing a task repeatedly. You spontaneously switch on and off your smartphone and do not have to look closely at the controls for doing so. Similarly, you are so used to operating your computer every day that you can almost blindly switch on without the need to look at different controls. So,

FIGURE 7.9 Mental model.

due to repeated interaction with a device, one develops a mental model to perform a specific task. This is a dynamic mental model dealing with a mental model in task performance. A good example would be that when you wake in the night to go to the toilet in your house at, you are often able to do so without needing to switch on the lights! This is again due to the fact that one develops a mental model (dynamic) after moving about in one's own room for years.

Design pertinence. Static mental models are important in designing icons for user interfaces and for designing logos and symbols for public usage. Examples include the different icons on your mobile phone and personal computer. They also include symbols like fire, escape, slippery road in different public domains. A dynamic mental model involves designing device interfaces in such a way that the user is able to use his previous knowledge of using similar devices. So design consistency at the user interface such as the location of specific information and input controls should be consistent and not change. If a mental model (dynamic) is adhered to then the chances of making a mistake at the user interface are reduced.

7.7 STEREOTYPE

Stereotype is the way in which we expect certain things to happen. For example, the traffic system in London is always on the left-hand side and on the right-hand side in France. Switches in India are in the ON position while they are placed downward which is an OFF position in the United States. Depending upon the culture and the place we are born and brought up, our stereotype develops, so when you go from London to Paris then you might be in trouble for the initial few days on the road. It is possible that you may be hit by a car on the road while looking in the opposite direction for oncoming vehicles. In fact, people in some countries in Asia may accidentally keep switches of many electronic products in the ON position inadvertently. It's not their fault, it's made in France and the OFF position of the switch is in the downward direction and the ON in the upward direction. Normally, we read from left to the right. Many people in this world read from right to the left. This is also a stereotype of the users and hence stereotypes are also highly population-specific and at times are called the population stereotype (Figure 7.10).

Design pertinence. When products are manufactured in one country in Asia but are meant to be used in another country like the United States, it's very important that the population stereotype is adhered to. For example, if you design a product for the United States then you must ensure that the ON position of the switch should be in the upward direction and the OFF position in the downward direction. Similarly, if you are designing the graphics to depict the way to wash clothes on a detergent package and it is to be exported to the Middle East, then the horizontal graphics should read from right to left and not the other way around as it might then mean something totally irrelevant. For example, this could become serious when combat aircraft are manufactured in one country and exported to another where people read from right to left. There, reading the display inside the cockpit in the opposite orientation may prove to be fatal for the pilots. In this case, the control should be customized and

FIGURE 7.10 Stereotype. (Photo by Little Visuals from Pexels.)

redesigned as per the population stereotype or the users should be given adequate training to operate such an aircraft.

7.8 COMPATIBILITY

This is the relationship between the stimulus and the response it produces. If the response is as per the user's expectation then the stimulus response relationship is said to be compatible. If you turn the latch of your door it opens (Figure 7.11) and that's natural, you expect that. If you tap on a table with your pencil you expect a sound to be generated. This is as per your expectation and hence it's compatible. On the other hand, if you tap on your table with a pencil and the door of your room suddenly opens, this is not what you expected and hence it's not compatible, or is incompatible.

Design pertinence. Rotating the steering wheel of your car in the clockwise direction turns the car to the right, and anticlockwise the reverse direction. These two movements are compatible. When you use your television remote control to navigate between channels, then the placement and the shape of the remote controls should be such that navigating between channels becomes easier. For example, the increment in channel numbers could be associated by placing controls with an upper arrow and decrement with a lower arrow. Compatibility ensures that errors are reduced and the users learn to use the device quickly with greater satisfaction.

FIGURE 7.11 Compatibility. (Photo by Pixabay from Pexels.)

7.9 COGNITIVE ERGONOMICS IN ISOLATION?

The discussion above should not give you the wrong impression that cognitive ergonomics is something in isolation. In fact, cognitive ergonomics and physical ergonomics are intricately associated with one another. The best example for this could be the use of a hammer by a carpenter. When the carpenter initially drives a nail into a wooden plank he first aligns the tip of the nail with the plank, then he applies some light strokes on the head of the nail by the hammer. When a portion of the nail tip enters the plank he gradually applies more force on the nail head using the hammer. When the nail is through, he exerts maximum force to ensure that the nail head is flush with the wooden plank. In this example you see that a simple task like hammering requires careful manipulation of the hammer, which leads to an increase or decrease of force as and when required. This manipulation happens from the brain and is a part of cognitive ergonomics. Experienced carpenters know exactly when and how to change this force and in what manner. Thus, cognitive and physical ergonomics are associated and not dissociated as humans think (Figure 7.12).

7.10 COGNITIVE ERGONOMICS ATTRIBUTES AND GLOBAL VARIANCE

The main principles of cognitive ergonomics remain constant throughout the world. Major variations are noticed in terms of stereotype, compatibility, and mental model.

FIGURE 7.12 Cognitive ergonomics in using a hammer and nail.

These have extreme design pertinence given the fact that the world is now a small global village with products being conceptualized at one corner, manufactured at another, and used in yet another part of the world. Hence the product that we design is a frequent traveler.

7.11 ASSIGNMENTS

1. Pick up an air-conditioner remote control and list the cognitive ergonomic issues in it.
2. Select any mobile phone of your choice and list the issues related to stereotype and compatibility.
3. While operating the microwave oven at home, identify the different stages of information processing and what possible design intervention could be done at each stage.
4. See how you can incorporate certain ergonomic features in digital products at home, such as the television remote control, oven controls, and your music player control and display so that the elderly are at ease using them.

8 Ergonomics of Display and Control

OVERVIEW

This chapter provides a walkthrough of the different ergonomic issues related to the design of display and control elements ranging from the mobile phone to the car. The elements of control and display compatibility are also explained and the different applications of cognitive ergonomics principles are also covered.

8.1 INTERFACE AND INTERACTION

These two terms, interface and interaction, are often confusing to students, and many make the mistake of using them interchangeably as if they are the same. Imagine that you are on a holiday on one of the most beautiful islands. There is another island adjacent to where you are staying and you wish to visit that island as well. Now what are the ways you can do it? You may take a boat and travel, or if there is a bridge then it's still easier and safer to travel even with loads of baggage. Only when you establish strong connectivity between the islands, either by building a good bridge or having reliable boat services, can you go to the other island and talk and interact with the people there. So, interaction with people means sharing your thoughts and feelings with them. To facilitate this interaction you need a strong bridge, which is your interface (Figure 8.1). So a strong interface (bridge) facilitates better interaction (establishing contacts with the people in the other island).

Displays and controls are the two channels in any human–machine system through which the user interacts with the technology. Imagine you are standing in front of an ATM, waiting to withdraw money for your weekend cinema visit. Your input to the machine goes through the keypad (control elements) as the machine has to understand what you are looking for. You cannot speak the machine's language, so the keyboard does the work for you. Similarly, the machine communicates to you in a language that you can understand through the display (output element). In essence, this forms a closed feedback loop wherein there are two points which are extremely vulnerable to errors. These two points are when you feed information to the machine through the controls and when the machine feeds information to you through the display. Ergonomics plays a vital role in strengthening these two vulnerable points and making the system robust, thus minimizing the chances of error at these two human–machine interfaces.

FIGURE 8.1 Difference between interface and interaction.

8.2 CREATING THE BASE FOR DISPLAY AND CONTROL ERGONOMICS

To understand the ergonomic issues in designing controls and displays, some of the fundamental principles need to be clear. First, the systems concept that we learned in Chapter 2 is very important. All display and control design exists in a context with the users and their diversity, the environment and the product/space or communication in question. You also need to define the criteria on the basis of which the different functions of the systems are divided between the users and the machine (called the allocation of function). One needs to consider the different technologies

FIGURE 8.2 Systems perspective of control and display design. (Photo by Pixabay from Pexels.)

and the implications of design issues in the selection of different controls and displays. Thus, to design an effective control and display from an ergonomic perspective we may follow some of the steps mentioned below (Figure 8.2).

8.2.1 Breaking Your Task into Smaller Components

If you break up the task that you want your system to perform into smaller tasks, it becomes easier for you to analyze it in depth. For example, you are asked to break up the task of brushing your teeth. If you intend to break it up into smaller elements they would look like: open the tap, wet the brush, apply paste, start brushing, rinse your mouth, and your brushing is over. It becomes much easier to look into the entire task of brushing your teeth in finer detail. Now you can see and take a decision about which stages need a human operator and which stages could be replaced by a machine (if it needs replacement).

This is based on what do you want to achieve (in this case, brushing your teeth) and the characteristics of the users (young people, elderly with trembling hands, people with stiff hand joints, etc.). This analysis would also tell you how and where a human will interact with the system, and ensure that the human does not become a weak link in the system (Figure 8.3).

8.2.2 Dividing Responsibilities between Humans and Machines

We have already seen different human characteristics. Similarly, machines also have certain characteristics of their own. For example, machines are good at repetitive,

FIGURE 8.3 Task analysis simplified.

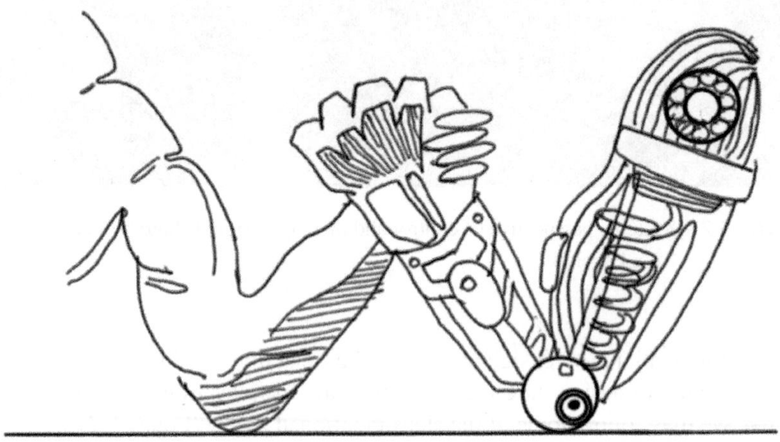

FIGURE 8.4 Allocation of function and the stages.

complex, and long calculations, assisting humans in keeping a vigil in any task, can perform tasks in parallel to humans, and so on. Thus, based on these two sets of characters (human and machine), one needs to decide at the user interface what responsibilities (function) could be given to the human and what could be given to the machine. It's similar to dividing responsibility between human and machine at the junction or interface between human and machine. This is also called "allocation of function" (Figure 8.4). You have been given the responsibility of organizing a cultural event at college. You cannot organize the event on your own. So, you call your friends and divide the responsibilities among them. One takes charge of the stage decorations, the other takes charge of catering, and the others take charge of transportation and other logistics. This division of responsibility between man and machine at a system user interface is allocation of function.

8.2.3 Stages in Dividing the Responsibilities between Humans and Machines

i. If the main objective of your design has an element of uncertainty, e.g., you want to send your girlfriend a cup of hot chocolate (crazy idea!) in the early morning of Valentine's Day just to surprise her. If you use a drone then you have to be double sure that the drone will not drop the hot chocolate on her face! As this task has some risk involved, you give its overall charge to a human, who could at least monitor the movement of the drone and ensure safe delivery of the product. If a drone is sent on this mission or an automated system is setup to do this task, then in case the action needs to be reverted at the very last moment (your friend might be sleeping) it is easier if there is a human who can make the necessary decisions.

ii. Humans make mistakes when they are made to do too many things at a time. If you are made to play the guitar, read a book, and watch television all at the same time, you know what's going to happen! At this juncture, machine or automation is preferred.

iii. Whenever there are too many different types of humans working at a system it invites different variations because all humans are different (Chapter 1). In these cases where different humans work at different parts of the system, they become a weak link in the overall system, because humans are different and they each have their own way of doing a task. If the main door of your house is made up of one single piece of wood it is much stronger compared to a door made of 20 pieces held together by small metallic strips! In such cases, a backup human operator is necessary to intervene if anything goes wrong or is about to go wrong. After all, humans can manage humans better than machines can!

iv. Even where automated systems are used, human operators should always be used as they are reliable and can take instantaneous prudent decisions depending on the circumstances if anything goes wrong. So trust humans!

v. Depending on the context, human beings need to be given complete charge of the system, where an instantaneous decision needs to be taken. For example, inside the aircraft if anything goes wrong then the pilot has to decide what best can be done, a machine cannot do that.

8.2.4 The Context of Use

Every product is used in a particular context which dictates the design of display and control elements. For example, the LCD screen of a smartphone might be legible indoors but in bright sunlight it becomes completely illegible (Figure 8.5). Therefore, the design of control and display elements is dictated by three things: users and their characteristics, the physical environment and the context in which it will work, and the tasks that we intend to accomplish with the product. Users' characteristics deal with different user profiles such as age, gender, anthropometric dimensions (Chapter 3), strength, cognitive function, social belief, knowledge level, stress level, and lifestyle. In general, there are differences among male and female users in terms of anthropometric dimensions of different body parts and also in terms of strength, where females have relatively lesser strength than their male counterparts. The dress and lifestyle of females being different from males also needs consideration. For example, females have longer nails and this might interfere with designing certain products such as

FIGURE 8.5 Context of usage of a product.

controls inside cars, where additional space has to be provided for longer nails. Similarly, special considerations have to be provided for senior citizens whose anthropometric dimensions are different and whose cognitive function may be slower compared to other citizens.

The physical environment includes heat, light, humidity, and the noise level of the environment. This also entails the type of posture and dress worn. It includes different conditions such as rain, earthquakes, floods, and also conditions such as perspiring, shivering, and so on. When designing a kitchen tap you need to consider the fact that the hand will be wet, possibly coated in oil or soap. Thus, the shape should be such that the hand under this condition is able to open and close the tap. In winter, people wear gloves and extra clothing. Thus, the design of cars and buses should factor in extra spaces, as one cannot have a customized space for different seasons.

Tasks normally refer to the duration, posture, strength required, grip or torque or both, male and female or dual use, single person or multiple person usage, and space required for performing the task.

8.3 CONTROL DESIGN

The following are some of the ergonomic guidelines in designing a control (Figure 8.6).

i. It should be visible to the eye, or should fall within the visual cone of the user. It is here that controls should be as per the expected area of the visual field. When you start your Windows PC every day at the office you are familiar with the date and time on the bottom right-hand side of the screen. As you see this every day you always expect it to be there. If by chance it shifts its position then you are uncomfortable. This is because that we develop a mental model and "expect" certain things in certain parts of the visual field. This expectation has to be respected when designing controls (as much as possible, because at some

FIGURE 8.6 Ergonomic principles in control design.

places controls also should be kept beyond the visual cone, e.g., seat adjustment control in a car). Unless the control is visible no action can take place.

ii. It should be accessible to the user. If the control is not accessible then there is no question of its activation. It should ensure that the body parts meant to activate them are in a position to exert maximum force. Try rotating and pulling a knob with your hands at the back!
iii. It should be identifiable. The user should know which control is meant for which specific function. To distinguish one control from the rest, color and shape or texture coding of the controls are done. At times, controls are made distinguishable by their position and relative size (a bigger circular button means more it is important compared to a smaller one) with respect to the display elements as well.
iv. It should be easy to use. The control should be easy to use by the user in the intended manner with optimum effort and hence issues such as body position, strength, dexterity, etc., need to be considered.

Along with the features mentioned above, all control designs should follow three principles, especially when considering the layout of the control elements at the user interface.

i. Most important first: Important controls should be placed in the central part and close to the body for easy access and activation.
ii. Frequency: Frequently used controls should always be within easy reach so that the body members do not fatigue easily, otherwise errors might happen. Controls used less frequently can be kept a little away from the body and reach.
iii. Sequence: It is important that controls be arranged as per the sequence of operation in tandem with the mental model in task performance of the users.

These three principles should be followed as and when applicable.

All controls should further look into the next principles to be applied as and when required.

i. *Using either hand*: In case of critical operations, controls should be placed in a manner that they can be operated by either hand. For example, in vehicles meant for military operations, if the human injures an arm, he should be able to bring his vehicle back to safety with the other arm.
ii. *No hindrance*: There should be enough clearance among the controls so that movement of one control does not lead to the accidental activation of another control.
iii. *Do not hide*: All controls should be visible as well as practicable. When a novice user is at the control panel, the controls should be within their visual cone. He should first learn their arrangement and develop a mental model of the same, then their activation will get easier.
iv. *Distribute input to the system*: If there are lots of tasks to be performed with the control, it's better to distribute the load for the two hands. This will ensure delayed onset of fatigue in the hands.

8.4 DISPLAY DESIGN

The sole purpose of a display is to convey information to the user in a format which is understood by them. Thus, we need to know the different modes in which information can be presented to the user, as information is what reduces uncertainty. The following different modes of information are available (Figure 8.7).

FIGURE 8.7 Different types of displays. (Photo by Maria Geller from Pexels.)

i. *Quantitative*: This is where the users would like to know, how much? For example, temperature, pressure, etc., where precise information is needed.
ii. *Qualitative*: The user here is interested in knowing the quality of the information, and not any precise information. Examples are quality of water, quality of rice in the market, etc.
iii. *Status information*: When information presents the status of the system, such as stop/go in traffic signals, on/off in lamps.
iv. *Alphabets, numbers and symbolic information*: Information such as traffic signs, labels on bottles, and instructions on washing machines fall under this category.
v. *Warning and signal information*: Here the absence or presence of information is key to what is being presented. As the name suggests, examples are warning signs (static and dynamic), the presence of which conveys information and the absence does not.
vi. *Representational information*: This is where information is displayed in the form of photographs, graphs, charts, to show the rate of change or the quantum and the nature of change.
vii. *Identification information*: These types of information help the user to identify hazards, traffic lanes, and electrical wires. Electrical cables are often color coded for easy identification. Many devices have color-coded sockets in them along with color-coded cables, which facilitate easy insertion of the cables into the sockets.

8.4.1 Display Design Principles

The following are some of the considerations when designing a display.

i. Characteristics of the task. The nature of the task will decide what manner the display has to be designed. If the task demands close monitoring of temperature change, then the display should depict the rate of change of temperature. If you are travelling in a car then the speedometer (display) should display your current speed. If you are frequently monitoring the speed of the car, the display should convey that information to you with ease. If the task is difficult, e.g., inside an aircraft cockpit, where the pilot has to monitor several displays at a time, then the design of the displays would guide what manner the information should be presented so that they are decoded easily by the pilot.
ii. Are you standing or sitting? It's very important that you are able to see the display. Thus, you need to keep in mind your eye height and visual cone (Chapters 5 and 6) while considering the placement of the display. When you are standing, the eye height is higher and the visual cone moves upward compared to when you are sitting down.
iii. Movement requirements in the task. At times, the user requires flexibility and movement in the workplace. The best example is a control room of a power plant where a few people (at times a single person) monitor different displays at the same time. In this case, the displays need to be designed not only in terms of their placement but also in terms of their content.

iv. Consider what gets in the way. Elements such as people and other displays can get in the way of the visual cone and hence need to be considered when placing the display devices. At times, the user has to look through obstacles, e.g., a driver looking at signage (static display) on the road through the windscreen of the car, which is covered by a thin film of water. In this case, the display design should be of sufficient contrast and size so that legibility is not impaired.

8.4.2 Use of Colors in Display

Colors make displays easy to understand, but one needs to use colors with prudence, otherwise the display could be confusing. The following are some of the ergonomic issues to keep in mind when dealing with colors in designing display elements.

i. Color in displays helps in better understanding.
ii. The human eye is most sensitive to the blue–green color spectrum; therefore to increase visibility, colors in this spectrum should be used.
iii. In Chapter 7 on cognitive ergonomics we discussed the dark and light adaptation of the human eye. This should be kept in mind when choosing colors, as people tend to adapt to certain colors, such as red in the dark and other colors in well-illuminated rooms.
iv. Pure red, pure green, and pure blue should be avoided because of the reasons discussed in Chapter 2. Doing so, will put color-blind people at ease with the displays that we design.

8.4.3 Other Display Types

Thus far, we have mainly discussed visual displays, where information from the system is only fed through the visual apparatus, the eyes. There are other types of displays as well, namely auditory (ear), tactile (touch), olfactory (smell), and gustatory (taste) displays. Auditory displays are mainly used for communicating in environments where the user cannot look at the visual displays, or where the illumination level is extremely low. These are also used in cases of people with low vision or who are visually impaired. Auditory displays also facilitate two-way communication between two users.

Tactile displays are mainly meant for the visually impaired. You will have seen these displays in the form of braille in ATMs, whereby through a mere touch of the hand they are able to understand the labels on the switches and the information is conveyed in an auditory/tactile mode.

Olfactory (smell) displays are where the display makes usage of our sense of smell. The LPG gas at your home has an odor which is deliberately added (LPG gas is odorless) so that you can detect any gas leakage. Some clocks are connected to specific perfumes at certain time of the day, so that the users in the vicinity get to know the correct time of the day just by smell (Chapter 7).

Gustatory displays or those that work with the taste attribute of the tongue are still in nascent stages. There is speculation that such displays would enable the user to receive information through the tongue, which if and when applicable would truly add to the multisensory experience of the users (Figure 8.8).

FIGURE 8.8 Different types of displays other than visual.

8.5 DESIGN OF A CONTROL PANEL

Designing a control panel plays an important role in display and control compatibility. Thus, the control and display elements need to be arranged with prudence so that the user's expectation is respected. There are three things which play a key role here in designing a good control panel.

i. *Physical ergonomics*: Where you need to decide upon the height of the panel with reference to different anatomical landmarks of the body. You should use the different principles in work space design that we discussed in Chapter 2. Next you need to decide upon the tilt of the panel (bigger panels are better seen if tilted). Open a newspaper. Keep it on two tables, one flat and one tilted. You can read all parts of the newspaper more effectively when it's placed on the table with the tilted surface because you can scan the elements effectively. The tilt of the control panel needs to be calculated from the user's eye position of so that they are able to see the entire control panel with ease. To calculate the surface area of the panel, the number and size of different display and control elements along with their different labels need to be considered. You should follow the principles of work surface design with elements aggregated into primary, secondary, and tertiary zones. This was discussed in the section on work space design in Chapter 3.

   A few ergonomic principles need to be followed in addition to those that we mentioned before. These are operational sequence, frequency of use, and center of attention, with important elements at the center. To ensure better compatibility controls should be placed close to the display and the placement of the controls should follow the logical sequence of movements of the display elements.
ii. *Good look and feel*: It's said handsome is what handsome does. Therefore, your control panel should look and feel good in order for the human operator

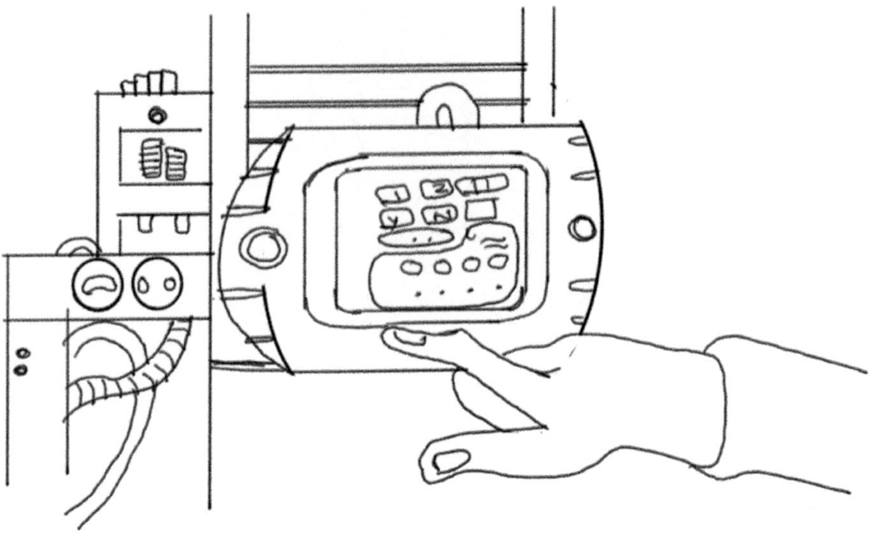

FIGURE 8.9 Designing ergonomic interfaces.

to enjoy his task. Graphics and labels on the control and display elements play a big role as well, as you need to design them based on the ambient illumination of the place and the distance from which they are to be viewed and controlled. These are discussed under visual ergonomics in Chapter 8.

8.6 DESIGNING ERGONOMIC INTERFACES

There are certain steps to be followed which may not be in the same sequence for all cases. First, you need to analyze the task to be done by breaking it into smaller elements or subtasks, as discussed before. They must be analyzed in depth in the light of different ergonomic principles of control and display design along with the different cognitive ergonomic principles discussed in Chapter 7. This should be followed by manufacturing feasibility of the product where you would need the help of an expert for selection of the right material and its fabrication. This is followed by designing the prototype (Figure 8.9).

8.7 ASSIGNMENTS

1. Go to a nearby ATM and identify the ergonomic issues in the controls and the displays.
2. Critically analyze your mobile phone; change the target user to an elderly user with reduced vision. Do an allocation of function at the device interface.
3. Design the control panel of a switchboard at home by incorporating all ergonomic issues in them.

9 Visual Ergonomics

OVERVIEW

This chapter provides an insight into the ergonomic issues in communication design pertaining to typography, icons, symbols, and the design of similar elements. The structure and function of the human eye with specific emphasis on the importance of the structure of the eyes in visual design are also discussed.

9.1 THE HUMAN EYE

To understand visual ergonomics you need to review the previous chapters on information processing, human senses, and cognitive ergonomics, as these are intricately related to visual ergonomics. This area of ergonomics is extremely important for communication designers who wish to design in both the print and digital media from the viewpoint of the users. We will discuss visual ergonomics here in the light of the different types of mediums that are used.

The structure and function of human eyes have been discussed. The eyes are in essence the gateways of the human body to the outside world through which information enters. They were essentially designed for distant vision and not the close vision for which we use them today. Thus, the problem with the visual apparatus lies in their misuse and not in their structure. This is the crux of all problems in visual ergonomics. We have the same vision today as always, but the task has changed. People use their eyes for close vision and at night, which the eye was not designed for. Hence, the system at the human eye–product interface is very weak and vulnerable to errors. As the eyes are the gateways to the world and to our mind, any stress to the eye is the biggest challenge for mankind today.

9.2 EVOLUTION OF COMMUNICATION MODALITIES

In pre–Industrial Revolution times, the mode of communication among humans was through gesture during the day time or when in close proximity, and through sound while at a distance or when the illumination level was too low. The eye was never used extensively for close vision, nor was the ear used for close loud music, as they are misused today. We live in an era where our sensory channels are grossly misused. The result is a mismatch between our sensory channels and the outside world, leading to human error of perception, which leads to systems failure. Constant near vision places a strain on the accommodation reflex of the eyes wherein the eye muscles fail to focus the lens to create a sharp image on the retina. Close audition leads to damage of the eardrum as well. There are specific movements of the eye called saccadic movements in which the eye actually moves from one word to the next (called saccades). These eye movements play an important role in detecting visual information from the outside world.

9.3 INFORMATION ERGONOMICS

Presenting information to users in any media is an art. It can be in different forms: visual (eye), auditory (ear), tactile or haptic (touch), and recently olfactory (nose) as well as gustatory (tongue). Because the channel capacity (the capacity of a sensory channel to detect, receive, and process information from the outside world) for eyes is the maximum, they play an important role in information processing.

9.4 APPROACH

In order to apply the principles of visual ergonomics in communication design we need to understand the following. The structure and function of the sensory modalities (the eyes in this case) are very important. Let us now divide the subject into two different sections: print and digital media.

9.5 PRINT MEDIA

This is also called static media. The selection of typography family is dictated by the quantum and nature of information to be presented to the users. In newspapers and tabloids, to save space information is presented in multiple columns with justification. Such justification of text creates unequal spacing (Figure 9.1) among words, and hence the specific movement of the eye (the saccadic movements) halts at the gaps and the reading speed is delayed. Therefore, to enhance readability in print media such as newspapers it is prudent to use left alignment of the text, so that the gaps between the words are uniform and the saccadic movements of the eyes progress uninterrupted. The reason why newspapers are displayed at an angle on a stand in many libraries is explained in Chapter 8. Upper- and lowercase typography is better comprehended by the eyes compared to using all capitals. This is because upper-/lowercase generate a pattern which the eyes can detect easily in less time compared to typography in all capitals. When it comes to all capitals, the entire word does not represent any pattern because all letters are of the same height and the outline of the same represents a square or rectangle. Thus, it is prudent to present information in upper- and lowercase for ease of reading and understanding. Only important information like DANGER or FIRE should be presented in all capitals so that they "shout out" in the crowd of information. Students are often confused about which is a better option in print media, a dark back ground and light text or a light background and dark text. The answer to this depends upon many factors. First is the ambient illumination. Next is the number of lines and words. The last is where you are going to place it for reading. If the level of illumination is optimal (let's say people are able to see and read properly) then a light background and dark text is preferable. If the level of illumination is low, for example on roads with very few lamp posts where it's difficult to read a newspaper, then a

FIGURE 9.1 Photo by Rawpixel.com from Pexels.

dark background and light text is preferable. If at such places you want to name the roads or house numbers and make them visible to users, then a dark background and light text is a better option. You need to exercise some caution while doing this in that the amount of textual material should not be too large or your users will have trouble reading them. Only the house number or the road name should be fine. The reason behind this is that under low illumination, the eyes are adapted to the dark (Chapter 7) and the pupil of the eyes are dilated (bigger). Hence, a dark background helps in further dilation of the pupil permitting more light. The light letters reflects more light and hence visibility even under low illumination increases. Thus a person is able to see well. If you use a large amount of text (four to five lines) on a dark background it becomes very difficult to read because the white text tends to create a feeling of "bleeding" or being spread in the dark background as the light text reflects too much light (as there are too many letters). Under optimal illumination, dark text on a light background is preferable.

9.5.1 Principles of Ergonomics of Text

The principles of ergonomics of any text in general follow three basic rules: communicating effectively, optimization of text organization, and economical text usage. The text needs to be organized first for which it needs to be aligned properly.

The type of alignment will be decided by the quantum of text. The third principle suggests the usage of minimal text, which has two advantages: users have less to read and hence less time is taken and less space is occupied on any surface. The third principle suggests the effectiveness of the text to communicate what is needed to the users, do not communicate or hide useless information. You need to keep in mind that humans are comfortable scanning images from top to bottom (the mechanism is complex and not that easy). Thus, when you see the complete photograph of your favorite hero or heroine you start by first looking at their face then gradually move downward and end at the footwear. Do you ever do it in the reverse direction? It can be done, but you feel more comfortable doing it from the head to the toe. Thus, when images or textual materials are placed one should keep this in mind, as the users would be at much more at ease in reading them. You can do a small experiment to test this out. Ask a small child to draw the face of a joker. What the child does first is draws a circle, then two small circles for the eyes, followed by the nose and the lips. Have you seen any child do it in the reverse direction? Seldom will you see someone put the lips in place, followed by the nose and ending with the eyes. This is because humans prefer scanning from top to bottom (although the mechanism is complex and there are issues of population stereotype associated with it).

9.5.2 Newspapers/Periodicals

For newspapers and tabloids one has to consider the ergonomic issues while placing a visual within the text and adding a caption to it as well. This is where one needs to follow the principles of control/display compatibility discussed in Chapter 8. The best compatibility is established with the display on top and the control just below it. The visuals are static display and the legends are like the controls, explaining the visuals. Thus, all legends should ideally go below the visual. If the space does not permit this, the second choice could be that the visuals are on the left and the legends on the right of the visual (depending upon the user's stereotype). One needs to exhibit caution while doing this. Cultural ergonomics suggests that the reading habit of users for text is normally from left to right (that's how English text is read). For other languages such as Urdu the reading takes place from right to left. So there is a change in stereotype of users in pattern of reading text as one changes geographical location. This needs to be respected or the users would not be able to interpret the text. When it comes to alignment, visuals on the top and text below the best option (Figure 9.2).

9.5.3 Packaging

For designing graphics on any packaging it's important to arrange the text and visuals judiciously within the small space available. The surface might be flat or it could be circular, textured, or any other shape. There could be variation of the material as well. The first ergonomic approach is to decide the exact area on the white space where users expect different information. Next, prioritizing the information

FIGURE 9.2 Ergonomic issues in text arrangement in newspapers and tabloids. (Photo by rawpixel.com from Pexels.)

is also necessary because some information is very important and should therefore stand apart from the rest. Apart from the principle that we discussed above, the preferred quadrant has to be tested. The central part of the visual field is the most prominent and hence important product information should be presented there. Normally, the brand or product name should be in a larger font size if it has to draw the attention of users from a distance. This is where the designer needs to know where exactly the product would be placed on the shelves in case of supermarket, the illumination level, and the nature of illumination. This would affect the color and the font used. Based on quadrant preference, the designer needs to chunk information for better recall and arrange it as per quadrants on the package. Let's take the example of a jar of jam. There are different types of users who buy jams. Some are inclined toward a particular brand and hence the information should only catch their attention from a distance. No other details are important or relevant for them. Users who prefer exploring different jams would like to know every detail of the jam such as ingredients, contents, and most importantly the price and expiry date. Hence ergonomics should take care of these elements. Any text is better scanned by the eye when placed horizontally than when placed vertically (when placed vertically, the number of lines increases and hence the reading continuity breaks). Hence, important and relevant information needs to be placed horizontally as much as possible. Other information which is important but cannot be placed horizontally can be placed vertically, but in that case the area needs to be highlighted for better

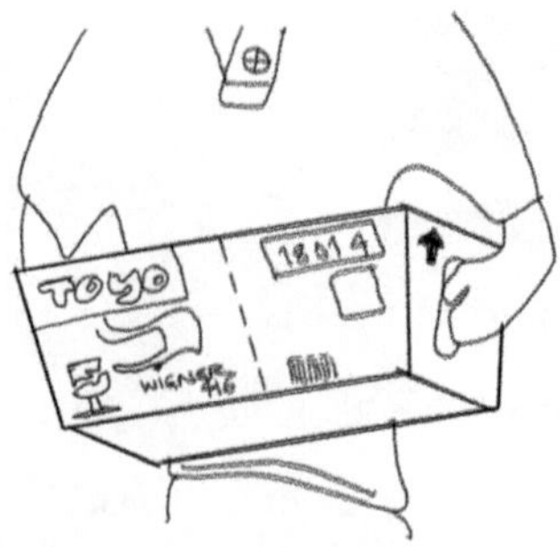

FIGURE 9.3 Ergonomic issues in packaging design.

sensation and perception by the human eyes. You need to keep in mind that in the assembly line different types of printers are used to print the price, expiry date, and manufacturing dates on products after all the packaging is over. Hence, the orientation has to be in tandem with the technology available at the assembly line. One has to study through task analysis (Chapters 2 and 4), to determine which part of the graphics will be covered by the human hand when holding the package. Therefore, the most important elements should be visible in such cases as well (Figure 9.3). The human eye has some characteristics which are very important when it comes to visual ergonomics. One of these is called "pattern recognition". The eyes are very good at recognizing any pattern. Thus, the names of many brands are available in a particular style, and people do not read the individual letter but recognize the brand the moment they look at the writing style or pattern. For example, a company launches a new chocolate in the market and names it "eat it"! Now let's say they decide to write it in this way as "Eat It". When people repeatedly see this style of writing, after some time people don't read them by the individual letters but see them as a pattern. Thus, such styles generate a very quick recall value among users because they see such writing as a complete pattern which becomes a mental model.

9.6 LOGOS/SYMBOLS/ICONS

Logos, symbols, and icons bank on the mental model of the users, which has been discussed before. Therefore, to increase the recall value of the users these elements should adhere to the mental model of the target users (Figure 9.4) or the mental model that the company wants to project to the users. Such projection of mental models when it relates to the real world is found in icons and symbols. Abstraction of mental models is found at times in the design of company logos. Too much detailing of these graphic elements increases the scanning time by the eyes and hence such elements are preferred to be simple without any detailed artwork. Colored graphics are better comprehended than black and white as eyes perceive everything in color much better compared to black and white. A touch of three-dimensional attributes to graphics makes recall higher as the human eye by stereotype perceives everything in three-dimensional perspectives. When

FIGURE 9.4 Ergonomic issues in symbol design. (Photo by Tim Mossholder from Pexels.)

designing such graphics for cross-cultural or international markets, it's important to adhere to the mental model of the target users, otherwise many of these will not be comprehended well or might be comprehended in the wrong manner.

9.7 WAYFINDING

This is also known as the "signage system" and is normally found in spaces to help users to navigate. The ergonomic attribute of navigating through any space necessitates three types of information. The first is how far is the destination (where I want to go), where I am right now, and how far I am from where I started (home). The other information that users might need is an exit route from the space. Thus, wayfinding ergonomics has to deal with this problem at many different levels (Figure 9.5).

1. Information. The quantum and nature of information is most important. The information should be short for easier user recall, as users will only be exposed to this information for a short period (Figure 9.5). The presentation of information, either through text and/or through visuals, has to be decided first. Next, the color and contrast and the size of the text needs to be decided. The text size would be guided by the distance from which it has to be read and the visual angle of the eye at that point. In a dark environment, a dark background with light text is required so that when the pupils dilate, they permit the entry of more light and increase visibility. If optimal illumination persists then a light background and dark text is preferable, as it facilitates reading because the white background reflects adequate light. These are the ergonomic rules of thumb and cannot be applied everywhere at random.

When designing information boards for external environments, it's essential that adequate contrast is maintained between the background and the foreground considering the long-term use of the same. For example, black text on a white background might initially appear to be an excellent contrast, but with exposure to solar radiation both black and white colors would fade away, and both tends toward a gray shade with reduced contrast. The same thing could happen when the information boards on the roadside are covered with dust. Therefore, one should select background and foreground colors in a manner in which they fade out in equal proportion so that the overall contrast between the two remains constant.

2. *Placement.* The exact placement of information so that it lies within the visual cone of the users is very important as well. Along with this, one has to decide the interval and at which strategic locations the information should be located. Information exposed to environmental hazards should factor in issues of visibility with the change in position of the sun. If the sun is just behind the wayfinding information then it will not be legible due to light adaptation (Chapter 7). To decide upon the height at which the information is to be located, one has to calculate the reference point from where the information needs to be viewed. For example, the visual cone of a person sitting inside a car and a truck would be different because the heights of the vehicles are different. Hence, the visual cone for people should be calculated (Chapter 5) with reference to their respective eye heights to decide upon the exact placement of the information. For vehicles which are high it's better to have the information above so that it's within the visual cone. For pedestrians the visual cone will be relatively lower and hence information should be attached to poles on the

FIGURE 9.5 Ergonomics of wayfinding. (Photo by Dominic M. Contreras from Pexels.)

side of the road on which the pedestrians move. The distance between the two sets of information is normally decided by the speed with which the users move. Normally, from a cognitive ergonomics perspective we have seen that first-time users look for reconfirmation in space about whether they are moving in the correct direction after every 5 minutes if they are on foot and almost the same time if they are in a vehicle. (The time duration is based on the author's own experience and is not based on any established scientific evidence.)

3. Break up of information. As we are poor in short-term memory, any information should be broken up into smaller components to aid users in steps giving small bits of information at a time. For example, if someone goes to a school building main gate looking for a specific teacher, how would you help them out with the information? If you have a list of all the teachers at the main gate with their respective room numbers, how would it be? This would be like searching for a needle in a haystack. So what options are there? You could arrange the names of all the teachers alphabetically, and in that case if the user knows the name of the person he could find it from the alphabetical listing. What if a user does not know the name of the teacher, but would like to meet the geography teacher? What would you do now? Would you give him a complete list of all the teachers? Not feasible. Hence as users are different, looking for different information, the challenge to us is to ergonomically design the information in such a way that it's effective for all. One ergonomic approach would be to arrange the names of the teachers in different departments. The second option would be to arrange the different names alphabetically. The third option would be to arrange the names of the teachers based on their location in the building. The most ergonomic approach would be to present information in smaller parcels as we are poor in short-term memory. So you have the first set of information with respect to different departments. The user is first guided to the department and does not need to remember anything else. While in the department, the user can now look for a specific name arranged alphabetically, as there is only a small chunk of information to be dealt with. Thus, the principles of ergonomics in information design, that is breaking up information into smaller chunks and presenting them as and when required and not all in one go works much better in such a context. There are different ways of presenting information. You can use color coding for shopping malls, numbers for different floors, or alphabets for specific zones. Sometimes a combination of all three acts as reconfirmation to the users as well.

4. Feedback. Users need to confirm whether they are moving in the right direction and hence feedback in the space is required to confirm the correct destination in the given direction they are moving. These feedbacks could come in the form of confirmation and reconfirmation of information on signage at regular intervals on the roadside. For example, you get out of the train station and have to check in to a hotel which you have been told is close to the station. Let's say that the name of the hotel is XYZ. As you move along the road, you find a board saying hotel XYZ is this way. You start moving in that direction. The signage appears at repeated intervals saying that "you are only _ meters from XYZ hotel". This is reconfirmation of information which keeps you going in the right direction until you come to the hotel, before which you again see a sign stating "you are almost here at XYZ hotel please turn left for the hotel reception".

5. Landmarks. We have seen that retrieval of information is easier with reference to landmarks, and hence we need to create landmarks so that this is facilitated. This could be in the form of a statue, fountain, or small garden on the road with a label so that the location of the target destination becomes easy. Your friend has invited to his house in the countryside. You are going to his house for the first time. You are told that because it is in the countryside there is not much signage on the road to guide you. So your friend tries to guide you in this way. He narrates "after getting off the bus at the Alpha bus station, take the road on your left. Walk for around 5 minutes till you come to a supermarket. Keeping the market on your right, take a right turn. After about 5 minutes you will find a 'surgery'. Right in front of the surgery you will notice a primary school. The school is next to my house". Here your friend has mentioned the names of a number of buildings such as the "supermarket", "surgery", "primary school", which are points of reference in the space (called landmarks), with reference to which it's easy for you to navigate and reach your destination. Thus landmarks in space are important as they help in the retrieval of information.

6. Bidirectional information systems. When you enter a large building and meet the person you want, often you are lost and cannot find your way back out of the building. Many public places are like that, they invite users inside, but once their task is complete they are left to themselves to get home! We are poor in short-term memory! So the exit needs another information set to show users the way out. These could be on the other side of the original information system so that you save on materials.

7. Information systems in space: Other ergonomic issues. The information system in public spaces should first catch the attention of users. For this it has to come within the visual cone, and should be in a contrasting color from the surroundings. If there are trees then the contrasting color of the information system should make it visible. Ergonomic principles suggest that the outdoor information systems should have a border in the same color as the text of the information, which would catch the attention and facilitate better scanning and detection by the human eye.

While using eye-catching colors in such contexts one should refrain from using more than three colors, otherwise it becomes confusing. While using colors pure red and pure green need to be avoided in the periphery as they are distracting while scanning. The preferred peripheral colors are yellow and blue. Chromatic colors can be used in large panels to attract attention of the users.

9.7.1 Five Ergonomic Principles of Symbol Design

The five principles are spatial compatibility, conceptual compatibility, physical representation, familiarity, and standardization. We have already discussed the principles of compatibility. These principles are mainly applicable when designing traffic symbols.

1. *Spatial compatibility*. This indicates the physical arrangement of the symbol in space relative to the information and the direction it is trying to convey. For

example, an arrow pointing to a building should point exactly at the building so that users understand it better.

2. *Conceptual compatibility.* This is the association between the symbol and the information it is trying to convey. It improves the user's comprehension and does not take long to decode it. For example, a person driving a car looking at the road symbol of a school ahead at once understands that they need to drive carefully.
3. *Physical representation.* As the word suggests, this indicates to what extent the road symbols represent the real world. For example, to indicate "Pedestrian Crossing", a symbol of a road with people crossing would be a depiction of the real world. This is physical representation.

9.8 ERGONOMICS IN CARTOGRAPHY (MAP DESIGN)

When a user searches for particular information on a map (Figure 9.6), the eyes start scanning and exhibit saccadic movements as discussed earlier. The situation can be compared with police trying to look for a criminal within a crowd. Your task becomes difficult if you have to locate a small place on a map, just like the police have a tough time when they try to locate a criminal with a cut mark on their right ear lobe. Ergonomic principles help to increase the efficiency of locating a place on a map. As the eye moves rapidly there is blurring of vision and hence the task is difficult. The total search time by the eyes is dependent upon the number of frames searched and

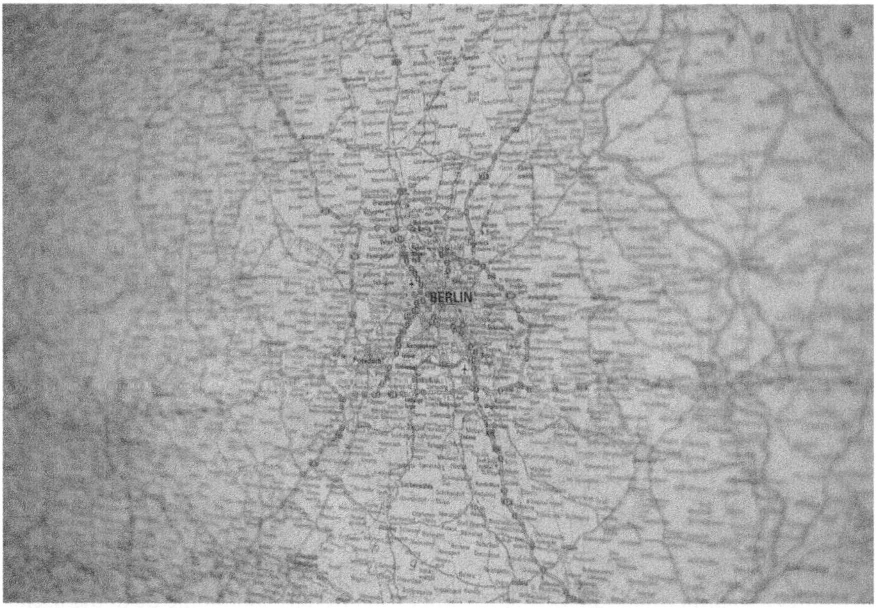

FIGURE 9.6 Ergonomics in cartography. (Photo by Pixabay from Pexels.)

the mean fixation time. The factors which affect fixation of the eye while scanning through a map are as follows.

1. *Number of names.* If the number of names is higher it increases the search time. Thus, in cartography one should try and eliminate redundant names.
2. *Typographic coding.* The selection of a specific family type also facilitates fixation. It has been found that names of places with initial capital letters are better fixated in less time.
3. *Color and size.* These are discriminated well in peripheral vision, but shape is not. Therefore, shapes should not be used to code places if peripheral vision is required.

It has been found that in maps the spacing between the letters conveys the area on the map. For example, if you write A N T A R C T I C A on the map of Antarctica the letters in capital (it could be initial capital and the rest in lowercase) which are spread apart with a space between indicates to the users that this name means a larger space as it spreads out all over the map. Similarly, if on the same map you want to indicate a smaller area called "humanity" (for example, there are no places by this name!) then the very placement of the closely packed letters indicates that it indicates a smaller area because the letters covers a relatively smaller area compared to Antarctica. In other words, these create a mental model among the target users and help them to decode the information easily on a map.

9.9 ERGONOMICS OF ICON DESIGN

Icons are the windows through which users interact with any multimedia devices. They can range from digital cameras and washing machines to cellular phones, all of which are heavily icon-dependent. It's through these devices that information is conveyed to the users. When icons are designed one should exploit the mental model of users. It has been seen that in many cases animated icons are better registered by the human brain if they are in tandem with the mental model of the users in task performance.

The ergonomic design of icons should take into account the user's mental model and the context in which the icons will be used to make it effective in conveying the desired information to the users. For example, when you are using an icon of a screwdriver to indicate "settings" on your touch-screen mobile, but when you are out in the sun this icon is no longer visible. So, when you change the context of usage the icon loses its meaning as it cannot be seen. Therefore, the success of any icon depends upon the characteristics (in terms of size, color, contrast, and placement) as well as how well users are able to understand them.

9.9.1 Ergonomic Principles in Icon Design

1. *Least space.* The icon should occupy minimal space on the screen or on the device surface and function appropriately as well as communicate to the users.
2. *Language independent.* Icon design should reflect the user's mental model and hence you can reduce the use of text at any multimedia interface.

3. *Increase recall value.* An ergonomically designed icon increases the recall value of the icon and thus a user's rate of scanning a page/surface on the multimedia device increases. This is significant because users in such situations skim rather than read all the information on the screen or surface.
4. *Styling quality.* This is affected by the message quality, meaningfulness, the power to locate it, and the judicious use of metaphors in icon design.

9.9.2 Ergonomic Issues in Icon Placement.

1. *Work in sets.* Icons are normally designed to work in groups or sets. Therefore, when designing the icon you need to keep in mind that it should fit the group and should not stand out as an odd man. This is where the size of the icon has to be calibrated relative to the other icons in the vicinity.
2. *Icon choice.* Your choice of icons should be guided by the context in which you are going to use it and your goal (what exactly you are trying to convey).
3. *Alphanumeric and pictorial icons.* Icons to be supplemented by alphabets or numerals or both will depend on how much information reconfirmation you are going to provide to the users. For example, in mobile phones many of the icons do not have high-recall value and require alphanumeric feedback when activated.
4. *Elderly.* As discussed, there are changes to the eyes as a user ages and hence icon design should adhere to this. The size of the icons should be large enough and the usage of pure blue should be restricted.

9.9.3 Cultural Ergonomics in Icon Design

Icons are a representation of the world around us and store information in an encoded form. Thus, designing icons needs to be done in tandem with the local culture and the way of thinking of the users. For example, a trash can is a product which most people are familiar with, as it's commonly used in offices and homes for throwing away useless things. This concept of a trash can is used in software to indicate that the files or folders not required can be thrown away. One need not write what it is as the image is a true representation of the real world. Unfortunately, if one travels to the east, there the concept of a trash can is unknown as people do not use it for throwing away their unused materials. So, the icon of the trash can when sold in the east would not be understood by the users and hence they would have trouble interacting with it. For these people, junk or unused materials are cleaned with a broom and thrown out directly without storing them. That's their culture. Thus, the trash can icon is better replaced with that of a broom in such a context.

9.10 ASSIGNMENTS

1. Select any text on a package and identify the visual ergonomic element in them.
2. Identify a space inside a shopping mall. With the help of ergonomic principles decide upon the placement and content of information elements in space.
3. Select a group of icons of a foreign digital camera. Identify the level of mismatch in a user's mental model for the local context.
4. Select a map of your town and identify the ergonomic issues in its design.

10 Ergonomics in Intangible Design

OVERVIEW

This chapter explains the intangible issues in design and the ergonomic reasons for the same. Issues such as the proper use of tools, the work–rest cycle, and the time of day effect and process design are explained in this chapter.

10.1 INTANGIBLE DESIGN

Designers in the domain of product and communication have an inherent tendency to become focused on the tangible aspects of design such as color, form, texture, and manufacturing. Rarely do they talk about the "other aspects" of design for which the design or the expected output from the design might not be satisfactory. Ergonomics in design becomes extremely important at this juncture. I am reminded of the story of our very dear general practitioner to whom we go for any small health issues ranging from the flue to complicated cases like a brain tumor. Let us take the case of someone suffering from high blood pressure or hypertension, as it is commonly known. The ergonomic approach to design is similar to the role of a general practitioner. The doctor gives the patient medicine for hypertension, but their job does not end here. After giving the medicine, they advise the patient to regularly exercise, to aim for a salt-free or salt-restricted diet, and to remain tension-free. This is because they know that to control hypertension medicine alone would not work, but other advice needs to be followed religiously along with it. So you can see that for the effective control of hypertension, a "holistic" perspective is necessary, and this is provided by the subject "ergonomics." The following sections explain in detail how ergonomics helps you in achieving that goal.

10.2 HAND TOOL/PRODUCT DESIGN

Even when the product is designed adhering to all design principles it might not give the desired benefit if the user is not taught about the proper usage of the product. You pick up a well-designed screwdriver. You use it for an unscrewing task (with the right hand) for 2 hours at a time. You are bound to suffer from forearm pain. Will you blame the designer for this? No. The fact is that our forearm muscle undergoes fatigue while unscrewing because of the structure of the forearm muscles when they move in an counterclockwise direction with the right hand. Here an ergonomist would come to the rescue and advise you not to perform this task for such a long duration, but take some rest in between. Another alternate solution could be to employ left-handed users, because their unscrewing activity would involve the stronger muscle group of the left hand and the fatigue onset would be delayed. Similarly, if one is using a vacuum cleaner to clean the ceiling of a house, they will suffer from shoulder pain after some time. Thus, the advice on the exact usage of the product is an ergonomic issue and should come from the designer otherwise he will be blamed for a bad design.

10.3 SEAT DESIGN

People always talk about an ergonomically designed chair, which is a misnomer. As we have seen, the structure of the body demands frequent breaks and changes of posture while seating, otherwise there will be pain and discomfort. Hence the designer has to tell the users this. The issue of back pain and psychological factors has been discussed, and thus one needs to be careful in dealing with back pain as a good backrest would not eliminate all types of back pain.

10.4 TIME OF DAY EFFECT

The different physiological parameters of the human body like heart rate, blood pressure, body temperature, rate of information processing, and other cognitive parameters are not the same throughout the day but exhibit variations. Even the strength and errors that a person makes vary throughout the day. This poses a serious challenge to designers today because products are used not at any particular time but as and when needed. For example, if the grip strength of a person is maximum at a certain time of the day and declines at some specific time, then the same product will not yield the same results even with the same person. Similarly, the interface of a mobile phone is used throughout the day, but vulnerability to errors is greater at specific times. The designer needs to keep this in mind while developing any product and test it on the users at different times of the day to ensure its functionality is not lost irrespective of the time of the day it is used.

10.5 GEOGRAPHICAL LOCATION AND DESIGN

The muscle fiber composition and the physiological cross section of the skeletal muscles dictate the strength of a person using a particular hand tool. Interestingly, both of these parameters of the skeletal muscle vary from one part of the world to another, and also within the same population. The design pertinence of this is that if a particular hand tool is designed keeping in mind the strength of the local people and is then used in some other countries or context for other people, then the tool might not give the desired benefit because the people of the country in which its being used might have a different muscle fiber composition and different muscle cross-sectional area, both of which determine the strength of a user. This is where ergonomics in design can be of immense help.

10.6 PREDICTING USERS FOR SPECIFIC TASKS WITH SPECIFIC TOOLS IN INDUSTRY

There are certain jobs such as heavy material handling, repetitive tasks, and precision tasks that some people perform. Not all of the users might be good at all these tasks at the same time. This is again dependent on the composition and structure of the human skeletal muscles and the energy expenditure of the human body. Therefore, while working with specific tools or gears in such tasks the designer needs to advise the management on the selection of specific users for specific tasks based on their physiological profile. Ergonomics can play a crucial role in selecting the right people for working with specific tools and machines for getting the maximum output.

10.7 ASSIGNMENTS

1. Select any hand-held product and identify the non-design issues in it.
2. What advice (intangible) would you give a furniture designer when designing a chair?
3. What intangible ergonomic features should you look for when designing the interface of an ATM?

11 Ergonomics in Color

OVERVIEW

This chapter introduces the reader to ergonomic issues when using color for products, spaces, and communication design. The ergonomic justifications for using different colors, as well as the reasons for using specific colors for specific areas in design, are explained.

11.1 BACKGROUND OF COLOR ERGONOMICS

We live in a colorful world. Just look at nature and you find so many different colors, and that's one of the reasons why people prefer staying in nature to enjoy their time. Just imagine that you are inside the office of a CEO of a multinational company. You notice that each wall of the room is colored with red, green, purple, and yellow. The ceiling is painted in a deep navy blue. What would be your first impression? You would probably think the CEO is eccentric! Why? It's because these colors do not provide that "corporate" look and feel. So color and its prudent usage in space are important. Here comes the importance of color ergonomics.

We have seen before that the perception of color by the human eye initiates at the retina and is completed in the brain. The key players behind color perception are the cone cells of the retina. The retina comprises three different types of cone cells: red, green, and blue. A mix of these different cells gives the impression of any color and is also referred to as trichromatic vision. Thus we are all able to see and enjoy the colorful world in which we live.

11.2 COLOR ERGONOMICS IN DIFFERENT DOMAINS OF DESIGN

If your bedroom is painted pitch black and your living room is painted deep navy blue, what can go wrong? Well, after you are home from the office, how would you feel? Irritated, bored? The colors are not in tandem with your mental state to make you feel relaxed.

Color in interior design changes the characteristics of the environment and is dominantly visible. In this aspect, light plays an important role in the visibility of color. Based on their wavelengths, colors are classified as cool and warm. Colors with shorter wavelength are treated as cool colors, like blue, whereas those with longer wavelengths are treated as warm, such as red. Therefore, rest rooms feel more relaxed and cool if painted light blue. Never paint your bedroom red but light blue, as that will give you a cool feeling and aids in relaxation. The visibility of any color has been found to be dependent upon the amount of light falling on it. Color in general has been found to increase attention span, develop a person's cognitive abilities, and refresh a user's perception toward the environment. Color can be used as a redundant cue for the retrieval of information in space. When you are navigating in a big building, color coding of certain areas like the waiting area, café, amenities and facilities such as the toilets and drinking water helps you to easily locate them in space.

11.3 COLOR BLINDNESS

We have seen how different categories of color-blind people (6% of males are red and green color blind, whereas less than 1% of females are affected) perceive different colors and how ergonomics can play a role in making these colors perceivable to a wider spectrum of the population. The ergonomic approach would be to make these colors impure with the use of additives in order to shift the spectrum and make them visible to the eye.

11.4 ERGONOMICS OF COLOR PERCEPTION

Studies have found that color in general affects human beings and these effects can be assessed through different assessments such as emotion, performance, and physiological parameters such as heart rate. Color has been found to impact human life and acts as a stimulus and affects humans physically, psychologically, physiologically, and sociologically every day.

Blue, green, and purple are considered cool colors as they have shorter wavelengths compared to red, orange, and yellow, which due to their longer wavelengths are considered to be warm. Many users perceive green as relaxing; which is possibly why actors wait in green rooms so that it relaxes them. For similar reasons, hospitals are painted in light blue or green colors. The color black signifies submission, which is why priests wear it as a mark of submission to the supreme.

The perception of color is not absolute and is affected by many factors. Illumination plays an important role in color perception as a change in illumination level leads to changes in the hue and saturation of the color. A user's response to different colors is guided completely by culture. In fact, preference for a particular color is also guided by the individual's understanding of the color term. The color red is perceived by some users as dangerous, some as auspicious, and by others as related to violence depending upon their prior experience. The perception of color also varies with gender as it evokes different emotions across genders. So as of now the concept of color perception is contextual and no conclusive evidence has been drawn. You can use color judiciously for making your product look attractive. For example, if you color the seating system in a shopping mall with green, then it becomes easy for users to identify it and also provides a nice soothing feeling.

11.5 INFLUENCE OF COLOR ON HUMAN PERFORMANCE

Color has been found to influence human performance to varying degrees when it is used for different types of designs. Appropriate color used for coding both online and offline forms are completed faster by target users with fewer errors compared to those which are not color coded. Color-coded controls and displays are better detected and comprehended and hence facilitate information processing in humans (Chapter 7). If you use color coding for indicating the level and type of "risk" or "danger" for any space or product then it is better understood. If you are in a building and see "Danger" written in red, you understand it much better and take it seriously compared to when the same word is written in green. You should refrain from using red and green together as it may lead to a blurring effect on the eyes. This is important if you are dealing with color in a space where manufacturing by human intervention is taking place.

11.6 COLOR ERGONOMICS AND DESIGN

Ergonomics of color usage cautions the designer to use colors prudently in products, space, and interactions as the effect of color on humans has been inconclusive to date. No color is good or bad, but its acceptance among users is dependent on many different factors like context, type of product, and a host of other factors described previously. Thus, while using color in design a detailed user study on the target population is warranted so as to get an estimate of the stereotype and the mental model of the users. As colors help to generate a specific mental model in users (Chapter 7) this needs to be incorporated when designing space and products. The meaning of the colors is highly context-specific and becomes acceptable as the context changes. So, although you might like the color red, you might not want to wear red trousers for an interview! Although black is a color for submission, we prefer black cars and they are used by the leaders of many countries as a symbol of elegance. Black leather shoes are used in any formal attire. Designers use colors to make certain products look attractive. You are drawn to a nice red refrigerator or to a pink smartphone and probably choose to buy it for its color. The preferred color makes you feel connected to the product and thus you buy it. For safety signs in the public domain try using a different color that has been found to be comprehended very well by the general public, including children.

11.7 ASSIGNMENTS

1. List the color ergonomic issues in a kitchen knife.
2. Discuss the ergonomic issues in designing for the red/green color-blind user when designing a signboard on the road directing users to a book fair in your city.

12 Exercises in Ergonomics Related to Design: Some Guidelines

CHAPTER 1: INTRODUCTION

Assignment. Select a pencil sharpener, a hammer, and a spectacle. Analyze them and list the good and bad ergonomic features in them.

Guidelines: Ergonomic features would address the issues of comfort, safety, and convenience. List them. To do this you need to hold each product and simulate its use in all possible ways.

CHAPTER 2: SYSTEMS PERSPECTIVE OF ERGONOMICS IN DESIGN

Assignment. Select a shopping mall of your choice. Try to identify the different subsystems and the connectivity between them. Identify the objective of the system and the point of vulnerability from an ergonomic perspective. How can you make this system robust with the help of ergonomic principles?

Guidelines: To solve this problem, first identify the objective of the system. Try listing the different subsystems, comprising different structural elements, shops, and people. Try connecting the different types of people with different elements of the mall such as the space, shops, security, amenities, etc. Identify the vulnerable areas in the entire system and suggest how they could be made more robust.

CHAPTER 3: HUMAN BODY DIMENSIONS

Assignment. Select a door handle and suggest what its exact dimension should be based on relevant anthropometric dimensions of the human body. You may use the anthropometric database for the same.

Guidelines: First identify the different touch points in the human body while one holds a door handle. In the second stage, try equating the different parts of the door handle with the respective anthropometric dimension. For example, the length of the handle is equal to the palm breadth. Ask yourself who are the target users, male, female, or both. If it is only male or only female then you need to look at the data for that specific gender. In the majority of cases the target users are both male and female. In that case you have to apply the principles of ergonomics. We have learned that for reach or access one has to start from the lower percentile value. For clearance, you have to start from the higher percentile value. So in cases of dual use by males and females for length of the handle you need to start from the higher percentile. Select the highest of the higher percentile value between males and females. Similarly, for diameter of the handle you have to start from the lower percentile value. Here also select the lowest of the values between males and female. Now you have a range of values. Start optimizing this value so as to satisfy the entire spectrum of the population.

CHAPTER 4: ERGONOMIC PRINCIPLES OF HAND-HELD PRODUCTS

Assignment. Select an agricultural or kitchen tool. List the ergonomic issues, focusing on the different ways people use them. Suggest a simple ergonomic solution for the same.

Guidelines: Ask a few users to use the tool. Observe them carefully and note the different ways it is used. Based on this observation, suggest your ergonomic solutions.

CHAPTER 5: ERGONOMICS OF SPACE

Assignment. Select a bathroom and a restaurant space. Analyze them from an ergonomic perspective and come up with two concepts for each. One should be within constraint without increasing the space but rearranging the elements in space and the other without constraint, wherein you can look into making the space very comfortable for users.

Guidelines: In ergonomic analysis of space you need to move in steps. First try visualizing the space in plan view (from the top as if you are a bird!) and then visualize it in elevation (from the side, front, or behind). The plan view will give you an insight into the work space requirement for users, what anthropometric dimensions to select, and how the space requirement can be minimized. The elevation view will give you insights into the aspects of height at which different elements need to be placed, what anthropometric dimensions to take. Then you need to do a task analysis of the space. This would involve having a plan of the space and mapping how people and material movement takes place to get an insight into the dynamicity requirement of the space. Crowd flow and group formation needs to be identified through user study. Once these analyses are done you can start optimizing the space requirement for individuals and groups of users based on different anthropometric dimensions. Along with this, design optimization of different furniture elements in space needs to be done based on anthropometric dimensions as indicated in other assignments.

CHAPTER 6: EXHIBITION ERGONOMICS

Assignment. Select a book fair in any open space. Apply the principle of ergonomics to ensure an organized flow of crowd and proper placement of all the books so that it becomes visible to all. Focus on specific user groups such as the elderly, physically challenged, and children.

Guidelines: The same principles followed in the case of space ergonomics are applicable here as well. The space would be decided by the number of people who are expected to be at the fair. For ensuring the crowd flow, the usage of barricades and information ergonomics would be critical. You need to decide on the location and number of amenities and facilities for the users based on the expected movement pattern of users.

CHAPTER 7: COGNITIVE ERGONOMICS

Assignment. Select the interface of a microwave oven at home. Perform a cognitive ergonomic analysis of the same and list the good and bad ergonomic issues.

Guidelines: This assignment can start with a questionnaire wherein you ask the user about the difficulty they face in operating the device. Then observe the user operating the device and note down your observations. Based on your observations, try to identify the cognitive ergonomic issues in each steps of operating the product.

CHAPTER 8: ERGONOMICS OF DISPLAY AND CONTROL

Assignment. Select the display and control of an ATM. Analyze it for ergonomics issues for users with low vision.

Guidelines: This assignment can be solved like the previous one. Conduct a task analysis, then start identifying the different ergonomic issues in display and control based on the ergonomic principles that you have studied.

CHAPTER 9: VISUAL ERGONOMICS

Assignment. Select a large shopping mall. Analyze the space for information system, symbols, and textual materials. Suggest an ergonomic solution for their improvement.

Guidelines: You need to have a good knowledge of the space and hence first get/generate a plan of the space. Do a crowd flow analysis, based on which you can suggest locations of information system, symbols, and textual materials.

CHAPTER 10: ERGONOMICS IN INTANGIBLE DESIGN

Assignment. Select any product or space and list the non-physical design issues in them. Indicate how they are related to the physical ergonomic component of the product or space in question.

Guidelines: Apply the principles of systems ergonomics to start with and then identify the ergonomic issues.

CHAPTER 11: ERGONOMICS IN COLOR

Assignment. A woman wishes to paint her two-bedroom apartment. The apartment has one living room, two small bedrooms, and one small balcony. She is an ophthalmic surgeon, who lives alone in this apartment. You have to guide her with proper color schemes from an ergonomic perspective for the entire apartment. The elements in her apartment range from one double-seated sofa set, one dining table, one center table, one open book shelf, one double bed, and one refrigerator. Use ergonomic principles in selecting your color schemes for the rooms and the elements in space.

Guidelines: Apply the principles of ergonomics in color and then give your suggestions based on that. She is a surgeon and hence leads a very stressful life. Your color schemes for the rooms should help her relax. Apply ergonomics of color to enhance the elements in space and make the space more vibrant for her.

Bibliography

Åstrand, P. O., Rodahl, K., Dahl, H. A., & Strømme, S. B. (2003). *Textbook of work physiology: Physiological bases of exercise*. Champaign, IL: Human Kinetics.

Balagué, F., Skovron, M. L., Nordin, M., Dutoit, G., Pol, L. R., & Waldburger, M. (1995). Low back pain in schoolchildren. A study of familial and psychological factors. *Spine*, 20, 1265–1270.

Baniya, R. R., Tetri, E., Virtanen, J., & Halonen, L. (2018). The effect of correlated colour temperature of lighting on thermal sensation and thermal comfort in a simulated indoor workplace. *Indoor and Built Environment*, 27, 308–316.

Bridger, R. (2008). *Introduction to ergonomics*. Boca Raton, FL: CRC Press.

Corlett, E. N. (1973). Human factors in the design of manufacturing systems. *Human Factors*, 15, 105–110.

Dianat, I., Haslegrave, C. M., & Stedmon, A. W. (2014). Design options for improving protective gloves for industrial assembly work. *Applied Ergonomics*, 45, 1208–1217.

Dreyfuss, H. (2003). *Designing for people*. New York, NY: Skyhorse Publishing.

Gamito, M., & da Silva, F. M. (2015). Color ergonomic function in urban chromatic plans. *Procedia Manufacturing*, 3, 5905–5911.

Gardiner, K., & Harrington, J. M. (Eds.). (2008). *Occupational hygiene*. Chichester: John Wiley & Sons.

Garner, S. (1991). *Human factors*. Oxford: Oxford University Press.

Guyton, A. C., & Hall, J. E. (2006). *Textbook of medical physiology*, 11th ed. Philadelphia, PA: W.B. Saunders.

Jones, L. A. (1917). The fundamental scale of pure hue and retinal sensibility to hue differences. *Journal of the Optical Society of America*, 1(2), 63–77.

Kaplan, M. (1995). The culture at work: Cultural ergonomics. *Ergonomics*, 38, 606–615.

Karwowski, W. (Ed.). (2005). *Handbook of standards and guidelines in ergonomics and human factors*. Boca Raton, FL: CRC Press.

Karmakar, S., Pal, M. S., Majumdar, D., & Majumdar, D. (2012). Application of digital human modeling and simulation for vision analysis of pilots in a jet aircraft: A case study. *Work*, 41(Suppl. 1), 3412–3418.

Karwowski, W. (2006). *International encyclopedia of ergonomics and human factors* (3 volume set). Boca Raton, FL: CRC Press.

Kroemer, A. D., & Kroemer, K. H. (2016). *Office ergonomics: Ease and efficiency at work*. Boca Raton, FL: CRC Press.

Kroemer, K. H. (2005). *'Extra-ordinary' ergonomics: How to accommodate small and big persons, the disabled and elderly, expectant mothers, and children* (Vol. 4). Boca Raton, FL: CRC Press.

Kroemer, K. H., & Kroemer, H. J. (1997). *Engineering physiology: Bases of human factors/ergonomics*. Chichester: John Wiley & Sons.

Kumar, S. (2001). Theories of musculoskeletal injury causation. *Ergonomics*, 44, 17–47.

Lehto, M. R., Landry, S. J., & Buck, J. (2007). *Introduction to human factors and ergonomics for engineers*. Boca Raton, FL: CRC Press.

Li, W. C., & Braithwaite, G. (2014, June). The investigation of pilots' eye scan patterns on the flight deck during an air-to-surface task. In *International Conference on Engineering Psychology and Cognitive Ergonomics* (pp. 325–334). Cham: Springer.

Lin, R., Lin, P. C., & Ko, K. J. (1999). A study of cognitive human factors in mascot design. *International Journal of Industrial Ergonomics*, 23, 107–122.

MacLeod, D. (1994). *The ergonomics edge: Improving safety, quality, and productivity*. Chichester: John Wiley & Sons.

Marras, W. S., & Karwowski, W. (2006). *Fundamentals and assessment tools for occupational ergonomics*. Boca Raton, FL: CRC Press.

McDougall, S. J., Curry, M. B., & de Bruijn, O. (2001). The effects of visual information on users' mental models: An evaluation of Pathfinder analysis as a measure of icon usability. *International Journal of Cognitive Ergonomics*, 5, 59–84.

Mukhopadhyay, P. (2006). Global ergonomics. *Ergonomics in Design*, 14(3), 4–35.

Mukhopadhyay, P. (2008). Time of day effect on performance at a user interface. *Multi: The Journal of Plurality and Diversity in Design*, 1(2), 35–44.

Mukhopadhyay, P. (2009). Ergonomic design of head gear for use by rural youths in summer. *Work*, 34, 431–438.

Mukhopadhyay, P. (2013). Ergonomic design intervention at an Interactive Voice Response (IVR) system in a developing country. *Information Design Journal*, 20(2).

Mukhopadhyay, P. (2013). Ergonomic design issues in icons used in digital cameras in India. *International Journal of Art, Culture and Design Technologies*, 3(2), 51–62.

Mukhopadhyay, P. (2017). Investigation of ergonomic risk factors in snacks manufacturing in central India: Ergonomics in unorganized sector. In *Handbook of research on human factors in contemporary workforce development* (pp. 425–449). Hershey, PA: IGI Global.

Mukhopadhyay, P. (2007). Heat stress in industry. *Asia Pacific Newsletter on Occupational Health and Safety*, 14, 12–13.

Mukhopadhyay, P., & Ghosal, S. (2008). Ergonomic design intervention in manual incense sticks manufacturing. *The Design Journal*, 11, 65–80.

Mukhopadhyay, P., & Khan, A. (2015). The evaluation of ergonomic risk factors among meat cutters working in Jabalpur, India. *International Journal of Occupational and Environmental Health*, 21, 192–198.

Mukhopadhyay, P., & Srivastava, S. (2010). Ergonomic design issues in some craft sectors of Jaipur. *The Design Journal*, 13, 99–124.

Mukhopadhyay, P., & Srivastava, S. (2010). Ergonomics risk factors in some craft sectors of Jaipur. *HFESA Journal, Ergonomics Australia*, 24, 4–14.

Mukhopadhyay, P., & Srivastava, S. (2010). Evaluating ergonomic risk factors in non-regulated stone carving units of Jaipur. *Work*, 35, 87–99.

Mukhopadhyay, P., & Vinzuda, V. (2019). Ergonomic design of a driver training simulator for rural India. In *Advanced methodologies and technologies in artificial intelligence, computer simulation, and human–computer interaction* (pp. 293–311). Hershey, PA: IGI Global.

Mukhopadhyay, P., Jhodkar, D., & Kumar, P. (2015). Ergonomic risk factors in bicycle repairing units at Jabalpur. *Work*, 51, 245–254.

Mukhopadhyay, P., Kaur, J., Kaur, L., Arvind, A., Kajabaje, M., Mann, J., … & Chakravarty, S. (2013). Ergonomic design analysis of some road signs in India. *Information Design Journal*, 20(3).

Mukhopadhyay, P., Vinzuda, V., Dombale, S., & Deshmukh, B. (2016). Ergonomic analysis and design of the console panel of a bus rapid transit system in a developing country. *The Design Journal*, 19, 565–583.

Mukhopadhyay, P., Vinzuda, V., Naik, S., Karthikeyan, V., & Kumar, P. (2014). Ergonomic analysis of a horse-drawn carriage used for a joy ride in India. *Journal of Human Ergology*, 43, 29–39.

Mukhopadhyay, P., Vinzuda, V., Sriram, R., & Doiphode, A. (2012). Ergonomic analysis of a traditional vehicle plying in rural and semi-urban areas in western India. *Journal of Human Ergology*, 41, 83–94.

Ng, A. W., & Chan, A. H. (2018). Color associations among designers and non-designers for common warning and operation concepts. *Applied Ergonomics*, 70, 18–25.

Panero, J., & Zelnik, M. (1979). *Human dimension & interior space: A source book of design reference standards*. New York, NY: Watson-Guptill.

Park, K. S. (2014). *Human reliability: Analysis, prediction, and prevention of human errors* (Vol. 7). Oxford: Elsevier.

Peters, G. A., & Peters, B. J. (2006). *Human error: Causes and control*. Boca Raton, FL: CRC Press.

Pheasant, S. (2014). *Bodyspace: Anthropometry, ergonomics and the design of work: anthropometry, ergonomics and the design of work*. Boca Raton, FL: CRC Press.

Pheasant, S., & O'Neill, D. (1975). Performance in gripping and turning – A study in hand/handle effectiveness. *Applied Ergonomics*, 6, 205–208.

Phillips, R. J. (1979). Why is lower case better? Some data from a search task. *Applied Ergonomics*, 10, 211–214.

Phillips, R. J., & Noyes, L. (1977). Searching for names in two city street maps. *Applied Ergonomics*, 8, 73–77.

Pincus, T., Burton, A. K., Vogel, S., & Field, A. P. (2002). A systematic review of psychological factors as predictors of chronicity/disability in prospective cohorts of low back pain. *Spine*, 27(5), E109–E120.

Prabir, M., Vipul, V., Krishnakumar, G., & Nirmay, S. (2017). Ergonomic evaluation and re-design of a camel-drawn public transport system. *Journal of Human Ergology*, 46(2), 47–58.

Prado-León, L. R. (2015). Color preferences in household appliances: Data for emotional design. *Procedia Manufacturing*, 3, 5707–5714.

Proctor, R. W., & Van Zandt, T. (2018). *Human factors in simple and complex systems*. Boca Raton, FL: CRC Press.

Sanders, M. S., & McCormick, E. J. (1987). Human factors in engineering and design. New York, NY: McGraw-Hill.

Siu, K. W. M., Lam, M. S., & Wong, Y. L. (2017). Children's choice: Color associations in children's safety sign design. *Applied Ergonomics*, 59, 56–64.

Wang, H., Liu, G., Hu, S., & Liu, C. (2018). Experimental investigation about thermal effect of colour on thermal sensation and comfort. *Energy and Buildings*, 173, 710–718.

Whittingham, R. (2004). *The blame machine: Why human error causes accidents*. Abingdon: Routledge.

Index

Note: Page numbers in *italic* denote figures.

T

U

V

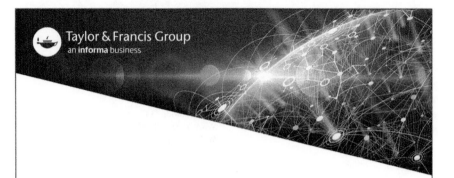